BENGAL DISTRICT GAZETTEERS

DARBHANGA

BENGAL DISTRICT GAZETTEERS

Burdwan/ **J.C.K. Paterson**
Darjeeling/ **L.S.S. O'Malley**
Howrah/ **L.S.S. O'Malley**
Hooghly/ **L.S.S. O'Malley**
Santal Parganas/ **L.S.S. O'Malley**
Sambalpur/ **L.S.S. O'Malley**
Puri/ **L.S.S. O'Malley**
Gaya/ **L.S.S. O'Malley**
24 - Parganas/ **L.S.S. O'Malley**
Purnea/ **L.S.S. O'Malley**
Bhagalpur/ **J. BYRNE**
Feudatory States of Orissa/ **L.E.B. Cobden-Ramsay**
Singhbhum, Saraikela and Kharsawan/ **L.S.S. O'Malley**
Cuttack/ **L.S.S. O'Malley**
Darbhanga/ **L.S.S. O'Malley**
Champaran/ **L.S.S. O'Malley**

BIHAR AND ORISSA DISTRICT GAZETTEERS

Patna/ **L.S.S. O'Malley**
Shahabad/ **L.S.S. O'Malley**
Monghyr/ **L.S.S. O'Malley**
Saran/ **L.S.S. O'Malley**

EASTERN BENGAL DISTRICT GAZETTEERS

Dacca/ **B.C. Allen**

BENGAL DISTRICT GAZETTEERS

DARBHANGA

L. S. S. O'MALLEY

LOGOS PRESS
New Delhi-110 002

ISBN-13: 978-81-7268-230-9

First Published 1907
Reprinted 2013

Printed and Published by
LOGOS PRESS
Building No. 4788-90, Street No. 23, Ansari Road, Darya Ganj,
New Delhi-110 002 (India)
Phone : 23272187
Email : mail@logospressindia.com
Website : www.logospressindia.com

PREFACE

WHEN the Statistical Account of Bengal by Sir W. W. Hunter was published, the district of Darbhangā had only recently been constituted, and it was treated as part of the old district of Trihut, in which it was comprised until1875. The present volume is, therefore, the first Gazetteer in which Darbhangā has been treated as a separate district.

I desire to acknowledge the very great assistance I have derived in compiling this volume from the Darbhangā Survey and Settlement Report (1904) by Mr. J.H. Kerr, I.C.S.

L. S. S. O' MALLEY

PLAN OF CONTENTS

TABLE OF CONTENTS

CHAPTER I

PHYSICAL ASPECTS

CHAPTER II

HISTORY

CHAPTER III

THE PEOPLE

CHAPTER IV

PUBLIC HEALTH

CHAPTER V

AGRICULTURE

CHAPTER VI

NATURAL CALAMITIES

CHAPTER VII

RENTS, WAGES AND PRICES

CHAPTER VIII

OCCUPATIONS, MANUFACTURES AND TRADE

CHAPTER IX

THE INDIGO INDUSTRY

CHAPTER X

MEANS OF COMMUNICATIONS

CHAPTER XI

LAND REVENUE ADMINISTRATION

CHAPTER XII

GENERAL ADMINISTRATION

CHAPTER XIII

LOCAL SELF-GOVERNMENT

CHAPTER XIV

EDUCATION

CHAPTER XV

GAZETTEER

GAZETTEER

OF THE

DARBHANGA DISTRICT.

CHAPTER I.

PHYSICAL ASPECTS.

THE district of Darbhangā, which forms the north-eastern portion of the Patna Division, lies between 25° 28′ and 26° 40′ north latitude and between 85° 31′ and 86° 44′ east longitude. It contains a total area of 3,348 square miles; and it is thus very much larger than any county in England or Ireland, except Yorkshire. Its shape is that of a fairly well-defined parallelogram, its mean length being greater than its mean breadth; its greatest length from north-east to south-west is 96 miles. The principal civil station, which is also the most populous town in the district, is Darbhangā, situated in 26° 10′ N. and 85° 54′ E. GENERAL DESCRIPTION.

The district is bounded on the north by Nepāl, on the east by the district of Bhāgalpur, on the west by Muzaffarpur, on the south-east by the district of Monghyr, and on the south-west by the Ganges, which divides it from the Patna district. Boundaries.

It takes its name from its chief town; and local patriotism insists that Darbhangā is derived from *Duārā Banga* or *Dar-i-Bangal*, meaning the door of Bengal. This derivation appears, however, to be philologically impossible and also historically incorrect, as the division between Bengal and Bihār has always been much further to the east. It seems more probable that the name is derived from that of the traditional founder of the town, one Darbhangī Khān, a Muhammadan freebooter, of whom little or nothing is known. The district formerly constituted part of Tirhut, a huge unwieldy district extending over 6,343 square miles; but this vast extent of territory being beyond the administrative capacity of a single Collector, it was divided into two in 1875, the western portion being constituted the district of Muzaffarpur, and the eastern and larger portion being formed Origin of name.

into a separate district, called Darbhangā after the name of its head-quarters. The name Tirhut is still, however, used as a convenient appellation for the country included in these two districts.

General configuration.

Darbhangā is a great alluvial plain with a general slope from north to south varied by a depression in the centre, corresponding roughly with the Wārisnagar thāna. The country lies on a low level, in many places indented with chains of shallow marshes, marking the lines of drainage by which the local rainfall and the overflow of the hill streams which intersect the district find their way southwards into the Ganges. The rivers flow on raised beds, which they have gradually constructed for themselves out of the silt brought down from the mountains in Nepāl. The alluvial plain, diversified only by these river ridges, is rich in all sorts of crops. In some tracts nothing but an enormous stretch of rice fields meets the eye, but in others the level plain is dotted with numerous clusters of bamboos and groves of mango and *sisu* trees.

Natural divisions.

Darbhangā is divided by its interior river system into three well-defined physical divisions. The first of these, starting from the south, is the tract in the extreme south-west of the district, comprising the thānas of Dalsingh Sarai and Samāstipur, which are separated from the rest of the district by the Little or Burh Gandak river. This area consists of a large block of upland with a few *chaurs* or marshes here and there, and is the richest and most fertile portion of the district. The second main physical division corresponds roughly with the Wārisnagar thāna and consists of a small marshy *doāb* between the Bāghmati and Burh Gandak rivers. This tract is the lowest part of the district, and is liable to inundation from the former river. The rest of the district, comprising the head-quarters and Madhubanī subdivisions, consists of a vast low-lying plain intersected by numerous streams and marshes, but traversed also in parts by upland ridges. The south-eastern portion, corresponding roughly with the thānas of Baherā and Ruserā, is, in the rainy season, mainly a vast chain of temporary lakes, joined together by the numerous beds of the hill streams which pass through the Madhubanī subdivision on their way from Nepāl to the Ganges. Large tracts in this area do not dry up till well on in the cold weather, and in some places communications are open for only three or four months of the year. In the Madhubanī subdivision, the land is generally higher, especially in the three western thānas and in the south of Phulparās, which contain stretches of high land.

RIVER SYSTEM.

The district thus contains three main river systems, the Ganges, the Little Gandak, and the Kamlā-Tiljūgā group of rivers. The Ganges forms part of the southern boundary of Darbhanga and further north the Little Gandak separates the greater portion of the Samāstipur subdivision from the rest of the district. That subdivision itself contains no rivers of any importance, but to the north there is a network of rivers and streams, called the Kamlā-Tiljūgā group, most of which converge on the extreme south-east corner of the district. In the Madhubanī and head-quarters subdivisions, practically all the rivers are liable to overflow their banks during heavy floods, but they rapidly drain off into the low-lying country in the south-east of the district, on which all the lines of drainage north of the Little Gandak converge. Their most marked characteristic in this portion of their course is that they flow on ridges elevated above the surrounding country, and each pair of rivers thus encloses a shallow depression, consisting of a series of *chaurs* or low lands leading into one another and forming the drainage channels of the country.

The Ganges. The Ganges skirts the district for 20 miles, but there are no large marts on its bank in this portion of its course. It is nowhere fordable at any time of the year, and its channel, when clear of sand-banks, is generally about a mile wide. In the rains, however, its breadth is much greater, large expanses of sand, which fringe its banks at other seasons, being then covered with water. These sand-banks are constantly changing, forming and re-forming in the most capricious way. Generally speaking, the bank on the Darbhangā side is sloping and ill-defined; and the low-lying lands in the neighbourhood are annually flooded.

The Bayā. The only stream of any importance which joins the Ganges direct, and not by way of the other river systems, is the Bayā. This river is an overflow of the Great Gandak, which forms the boundary between Muzaffarpur and Sāran. After flowing through the south of Muzaffarpur, it runs through part of the Dalsingh Sarai thāna in this district, and finally joins the Ganges just below Dhanespur at the extreme south-east corner of the Samāstipur subdivision.

The Little Gandak. The Little Gandak, or the Burh Gandak, as it is also called, is an important river throughout its course in Champāran, Muzaffarpur, Darbhangā and North Monghyr. In all these districts it marks a clearly defined division of the country. It enters the Darbhangā district near Pūsa, and after flowing past Samāstipur, leaves it just below Ruserā. Though its importance has been diminished by the railway, it is still a valuable trade highway, and there are many large bazars and marts on its banks. It is navigable, practically all the year round, for country boats of fair size. Its offshoots, the Jamwāri and the Balān, leave it near Pūsa and, after flowing through the south-west of the

Samāstipur subdivision, rejoin the parent stream in Monghyr, before it flows into the Ganges at Khagariā.

The Kamlā-Tiljūgā rivers.

All the rivers in the head-quarters and Madhubanī subdivisions belong to the Kamlā-Tiljūgā group, so called because they converge at Tilakeswar in the south-east corner of Ruserā thana, and are thenceforward known indiscriminately by either name, while proceeding through Monghyr and Bhāgalpur to join the Ganges and the Kosi by various tortuous courses. The first of the group, the Bāghmati, rises in Nepāl, and during its course through Darbhangā pursues an easterly direction parallel to the Burh Gandak; it formerly joined this river near Ruserā, but within the past 30 years it has cut out a new bed for itself, and now flows into the Karai and joins the Tiljūgā at Tilakeswar. The Karai, prior to its junction with the Bāghmati, is an unimportant stream. The Little Bāghmati, on which the town of Darbhangā stands, also finds its way to the Tiljūgā by the bed of the Karai. Its chief tributary is the Dhaus, which runs through the north-west of Benīpatī thāna. The Little Bāghmati was formerly joined near Kamtaul by the Kamlā, a river whose old beds are found all over the north of the Madhubanī subdivision. It used to flow 10 miles east of Madhubanī town, but now passes 10 miles to the west of it, its main channel running about 4 miles east of Darbhangā town past Baherā, Singiā and Hirni to Tilakeswar. It is a fairly large river in the rains and is liable to heavy floods. Still further east is the Little Balān, a deep and narrow river with a well-defined bed, which runs south through the eastern part of Khajaulī and Madhubanī thānas, and joins the Tiljūgā near Ruserā. The Balān proper, also known as the Bhāti Balān, is a river with a wide shifting sandy bed, also liable to heavy floods, but practically dry during a great part of the year. Its old beds are found all over the north of Phulparās thāna. Last comes the Tiljūgā, which rises in Nepāl and skirts the entire eastern boundary of the district, though parts of it lie in the Bhāgalpur district.

In Darbhangā and Muzaffarpur the Kamlā is worshipped as the younger sister of the Ganges, and receives similar offerings of goats, which are thrown alive into the river and are then taken by Mallāhs. Women pray to this river goddess for issue; children are taken to the river bank when their heads are first shaved; and she is worshipped when new tanks or wells are excavated.

GEOLOGY.

Darbhangā is a level plain rising very gradually towards the foot of the Himālayas, and with a belt of fairly high land along the bank of the Ganges. Between these two extremes the general elevation is lower, and considerable areas are liable to inundation. The soil consists mainly of the older alluvium or *bāngar*, a yellowish clay with frequent deposits of *kankar*, but in parts this has been cut away by the rivers rushing down from the Himālayas, and the low land, through which the latter find an exit to the Ganges, is composed of more recent deposits of sand and silt brought down by them when in flood. The soil of the district is thus entirely alluvial. It is impregnated in parts with saltpetre and other salts; and occasionally beds of *kankar*, or nodular limestone of an inferior quality, are met with.

"The essential features," writes Major D. Prain, I.M.S., in Bengal Plants, "of the vegetation in the area to the north of the Ganges, from the Gandak on the west to the Brahmaputra on the east, as we pass from north to south, are as follows. First, a narrow, more or less sloping, gravelly submontane tract along the base of the Himalaya, covered, except along river-beds, with a dense forest, the constituent species of which are those that occur on the lower slopes of the mountains themselves. In existing river beds only a few tough flexible bushes occur; along abandoned shingly river-courses the jungle is open and park-like, and the species are those characteristic of a drier climate than obtains in the forest alongside. This submontane forest is normally succeeded by a belt of swampy land of varying width, covered with long reedy grasses. Further out into the plain the ground as a rule rises somewhat, and, if so high as to be free from inundations, is in waste tracts usually covered with open jungle of a bushy character in the western parts, taller and more park-like in the central districts, and mixed with reedy grass or sometimes consisting only of tall grass as we pass to the east. Much of this tract, however, especially in the west, is under cultivation, and is then bare or diversified with bamboos, palms, and orchards of mangoes, or, less often, groves of other trees; in and about the villages themselves the mangoes are often accompanied by a number of tree-weeds and semi-spontaneous, more or less useful, bushes and trees." BOTANY.

For botanical purposes Darbhangā forms part of Tirhut, *i.e.*, the region lying from west to east between the Gandak and Kosi, and from north to south between the sub-Himālayan forest and the Ganges. The botanical features of this tract are in many ways different from those of Bihār, *i.e.*, the tract extending from the Son on the west to the old bed of the Bhāgīrathi on the east, and lying from north to south between the Ganges and the *ghāts* of Chotā Nāgpur. Together they form an integral portion of the Upper Gangetic plain, but, as Major Prain points out, "Tirhut is wholly flat, whereas Bihār is much diversified by hills Bihār, too, is appreciably drier than Tirhut, and these two circumstances, greater diversity of surface and less humidity, account for the presence in Bihār of many species that are absent from Tirhut. Another and, though an accidental, not less important factor in influencing the vegetation of Tirhut is the density of the population. So close, in consequence, is the tilth, that throughout whole districts field is conterminous with field, and the cultivated land abuts so closely on wayside and watercourse as to leave no foothold for those species that form the roadside hedges and fill the weedy waste places so characteristic of Lower Bengal. Even the village shrubberies that constitute so marked a feature of much of our area, are in Tirhut conspicuous by their absence. The result is that, except for the water-plants in the smaller streams and sluggish rivers, the vegetation of Tirhut is chiefly limited to the crops with their concomitant field-weeds; even the latter are often conspicuous by their paucity."

The following is an account of the different botanical species found in Darbhangā:—

The ground is under close cultivation, and besides the crops carries only a few field-weeds, except for a few very small patches of jungle, whereof the chief constituents are the red cotton tree (*Bombax malabaricum*), *khair* (*Acacia Catechu*) and *sisu* (*Dalbergia Sissoo*); in these parts there is an under-growth of euphorbiaceous and urticaceous shrubs and tree-weeds, like *Breynia*, *Trema*, *Flueggia*, *Phyllanthus* and *Glochidion*. Occasionally also large stretches of grass land are found, of which the chief species are *Eragrostis cynosuroides*, *Andropogon intermedium*, *Imperata arundinacea*, *Saccharum spontaneum*. These are interspersed with smaller spots of *ūsar* land sparingly beset with *Andropogon aciculatus*, *Diplachne*, *Sporobolus* and similar grasses. Near villages small shrubberies may be found containing mango, *sisu*, *Eugenia Jambolana*, various species of *Ficus*, an occasional tamarind and a few other semi-spontaneous and more or less useful species. Both the palmyra (*Borassus flabellifer*) and the *khajūr* (*Phœnix sylvestris*) occur planted and at times self-sown, but neither in great abundance. By the roadsides or round village enclosures, hedges of *Jatropha Curcas*, *Cæsalpinia sepiaria*, *Grewia* and similar shrubs are often covered with climbing species of *Convolvulaceæ*, *Tragia involucrata* and various species of *Vitis*. Hedge-row weeds are represented by *Jatropha gossypifolia*, *Martynia diandra* and similar plants. The field and roadside weeds include various grasses and sedges, chiefly species of *Panicum* and *Cyperus*; prostrate species of *Evolvulus*, *Indigofera*, *Ionidium*, *Desmodium*; and herbaceous species of *Phyllanthus*, *Euphorbia*, *Heliotropium*, and the like. In waste corners and on railway embankments thickets of *sisu*, derived both from seeds and root-suckers, very readily appear. The sluggish streams and ponds are filled with water weeds, submerged *Ceratophyllum*, *Hydrilla*, *Vallisneria*, *Ottelia*, and floating *Potamogeton*, *Nelumbium*, *Nymphœa*, *Trapa*, *Jussiœa*, *Ipomœa*, the sides being often fringed by reedy grasses and bulrushes occasionally intermixed with tamarisk bushes.

FAUNA. Tirhut was formerly famous for the variety of its fauna in the days when forests covered large stretches of land which are now under the plough. "The buffaloes," says the author of the Ain-i-Akbarī, "are so savage that they will attack a tiger. There are many lakes, and in one of them the water never decreases, and the depth is unfathomable. Groves of mango trees extend to a distance of 20 *kos*, delighting the eye. In the rainy season gazelle, deer and tiger frequent the cultivated spots and are hunted by the inhabitants. The deer they surround with an enclosure, and take them when they please." Even towards the close of the 18th century wild animals were still very plentiful. A few years before the Permanent Settlement rewards were paid for the slaughter of 51 tigers in a single year; there was much waste land in the north of the district inhabited by the beasts of prey that cultivation had driven out of the more southern tracts; and the depredations of herds of wild elephants were a serious danger.

The advance of cultivation, the growth of the population, and the extension of means of communication have now driven the wild animals which formerly infested the district back to the jungles of Nepāl; and there is no jungle left sufficiently large and dense to provide shelter for the larger beasts of prey. Leopards are occasionally found in patches of jungle towards the north, but only for a few months at a time, and they are merely stray visitors from Nepāl. Wolves and hyænas are met with but rarely. Jackals are numerous; they often carry off goats and kids, and sometimes children. The fox, wild cat, martin and civet are the other small predatory animals common in the district. The mongoose, hare and porcupine are plentiful towards the south. The only representatives of the Ungulata are *nīlgai* (*Boselaphus tragocamelus*) and wild pig. The former are not common, but are sometimes found in considerable numbers in large patches of thatching grass. Lately, however, there has been an influx of *nīlgai* causing considerable damage to the standing crops. The wild pig has also become scarcer, as the waste lands and patches of jungle and grass which were formerly common have been brought more and more under cultivation. He does little damage, except in the immediate neighbourhood of a few large grasses, where he now makes his home. Whenever he can be found, he is still hunted with the same keenness by the Europeans of the district; and though his kind has decreased in number, he retains the same qualities of cunning, hardihood and courage which make pig-sticking one of the finest sports in the world.

Game birds.

The game birds of the district are not numerous. They consist of quail of various kinds in small numbers, a few snipe, grey partridge, and a few black partridge in patches of thatching grass. Duck of all kinds come in at the beginning of cold weather. They are found in the early part of the season in the big *chaurs*, where rice is growing, before the water has dried up. They go later to the numerous tanks which are found all over the country. The following kinds of duck and teal are known:—the red-headed, white-eyed and crested pochard, pin-tail and shoveller duck, widgeon, ruddy sheldrake, blue-winged, whistling and cotton teal. The spotted bill duck, whistler and cotton teal breed here. Snipe, plover, cranes, storks, curlew and numerous sorts of waders are regular visitors.

Fish.

Most of the rivers and lakes, and many of the tanks abound in small fish, such as *rahu*, *nuni*, *jasīr*, *bachuā*, *tengrā*, *singhī*, *katlā*, mullet, a great variety of the smaller carp, poach and dace-like fish, a predatory fish known as *boāri*, and a flatsided fish called *buna*, some of which grow to a large size. The *hilsa* has sometimes been found in the Little Gandak. In the latter river and in the Bāghmati and other larger rivers, the Gangetic porpoise is fairly common. The common turtle is also found in the larger rivers and lakes. The *ghariāl* or fish-eating crocodile and the mugger or snub-nosed variety, locally called *boch*, are found in most of the rivers. The former are occasionally of great size, *ghariāls* 20 feet and more in length having been shot.*

CLIMATE.

The climate of Darbhangā is generally dry, bracing and healthy. The range of the thermometer is not so great as in the south of the Patna Division; and though the heat is greater than in the deltaic districts of Bengal proper, the dryness of the atmosphere renders its effects less enervating. The year is divided into three well-defined periods—the cold weather, the hot weather, and the rainy season. The first commences early in November, and may be said to last till nearly the end of March; for though the days then begin to be hot, the nights and early mornings continue comparatively cool and fresh. It would be difficult to find a more delightful climate than that which Darbhangā enjoys at this time of the year. The days are bright, warm and invigorating; and as soon as the sun sets, the temperature falls, and a

* I am indebted to Mr. W. Egerton, I. C. S., Collector of Darbhangā, for this account of the Fauna of the district.

fire is at once a comfort and a necessity. The hot season, commencing early in April, is ushered in by dust-storms and west winds, which often cause the temperature to rise above 96° in the shade. The heat is most intense in May, and unless broken by frequent rain, continues to increase until the gathering clouds herald the approach of the rainy season. In an exceptional year, the wind may blow steadily from the east, but such seasons, though cool, are not healthy. The west winds, on the other hand, though they parch up vegetation and raise immense clouds of dust, are generally healthy, and even pleasant, for when they blow, the interior of the houses can be kept cool by means of screens of moistened scented grass (*khas-khas tattis*); these are placed at the windows, and kept constantly wet, so that the hot wind blows into the house cooled and tempered. About the middle of June is the usual date for the commencement of the rains. Sometimes a few days' downpour is succeeded by a week or ten days of fair weather before the rains really break. The rainy season continues till the end of September or the middle of October, when the saturated ground begins to dry. This is considered to be the most trying season of the year; the air is hot and steamy, while the sun's rays are still powerful. The temperature then gradually diminishes, and by November the mornings become perceptibly cool.

Temperature and humidity. In spite of the dry westerly winds experienced in the hot season, the temperature is not so excessive as in South Bihār, the highest on record at Darbhangā being 107° in 1894. The mean maximum temperature ranges from 73° in January to 96° in April and May, and falls to 75° in December, more than half the change taking place in November and December. The mean minimum temperature varies from 52° in January to 80° in July, the lowest ever recorded being 38° in January 1878.

Humidity is, on an average, just over 60 per cent. of saturation in March and April, 70 in May, and generally between 80 and 90 in other months, the highest being 88 in July and August.

Rainfall. During the months November to April fine dry weather, with an almost entire absence of cloud and rainfall, prevails, and only a fraction of an inch falls monthly. In normal years the monsoon breaks in June, when there are 7·4 inches of rain; the heaviest rainfall occurs in July and August, varying from 12·5 inches in the former to 12·9 inches in the latter month; and the rainfall then decreases to 9·9 inches in September. At this period rainfall is heavy in the submontane tract, partly owing to the heavy showers which occur when cyclonic storms break up on reaching the hills, and partly because the monsoon current is stronger towards the west along the foot of the hills.

Statistics of the rainfall at the various recording stations are given below for the cold weather (November to February), the hot weather (March to May), and the rainy season (June to October). The figures shewn are the averages recorded from the earliest year in which rainfall was systematically registered up to the end of 1906.

STATIONS.	Years recorded.	November to February.	March to May.	June to October.	Annual average.
DARBHANGA	30—31	1·28	3·65	44·16	49·09
BAHERA	15—16	1·39	4·22	48·56	54·17
RUSERA	15—16	1·28	3·11	44·58	48·97
MADHUBANI	28—30	1·36	4·21	44·53	50·10
SAMASTIPUR	30—31	1·41	2·93	42·17	46·51

CHAPTER II.

HISTORY.

PRE-HISTORIC PERIOD.

DARBHANGA, lying on one of the great lines of Aryan immigration, was occupied at an early period by races of Aryan descent; and according to the legend preserved in Vedic literature, it formed part of the country in which the Videhas settled on their migration from the Punjab. The legend relates that Agni, the god of fire, accompanied the Videhas on their march eastwards from the banks of the Saraswatī, and when they came to the broad stream of the Gandak, informed them that their home lay to the east of the river. Thenceforward, the Videhas lived to the east of the Gandak, where they cleared the marshes, cultivated the virgin soil, and founded a great and powerful kingdom. This kingdom was in course of time ruled over by king Janaka, round whose name a halo of legend clings. Under his rule, according to Hindu mythology, the kingdom of Mithilā was the most civilized kingdom in India. His court was a centre of learning and attracted all the most learned men of the time; Vedic literature was enriched by the studies of the scholars who flocked there; his chief priest, Yājnavalkya, inaugurated the stupendous task of revising the Yajur Vedas; and the speculations of the monarch himself, enshrined in the sacred works called the Upanishads, are still cherished with veneration by the Hindu community. The centre of this enlightened kingdom of Videha was apparently in Tirhut; the capital was Mithilā, which is probably identifiable with that of Janakpur, a town a short distance to the north-west of the district in Nepāl territory, the name of which preserves the memory of the scholar-king who made it his capital; and tradition points to the village of Phulhar in the north-east corner of the Benīpatī thāna as the flower-garden where the king's priests used to gather flowers for the worship of the gods, and identifies its temple with that of the Devī Girijā worshipped by Sīta before her marriage with Rāma. The legends inspired by local patriotism make this tract of country the home of several *rishis*. At the junction of the rivers Jamunā and Kamlā there dwelt the great sage, Jai Muni Rishi, who compiled a learned treatise on the laws of justice in relation to the worship of the gods; the village of Kakraul at the eastern junction of the Kamlā and Karai was the abode of the sage Kapildeva Muni, the celebrated author of the collection of Vedic hymns called the Sānkhya Yoga; while in the north-west corner of the Darbhangā thāna there stands to this day the Brahmanical village of Ahiāri, where lived the third member of the famous trio of Tirhut sages, Gautama Rishi, the author of the Nyāya Sāstra and the husband of Ahalyā, from whom the village is said to have derived its name. Similarly, the neighbouring village of Bisaul is believed to mark the site of the hermitage of the sage Viswatra, and at Jagban, near the Kamtaul railway station, a banyan tree extending over a large area is adored as the hermitage of Yājnavalkya.

EARLY HISTORIC PERIOD.

The earliest event, however, which can claim historic reality is the rise of the Vrijjian oligarchical republic, which apparently replaced the old monarchical rule of Videha, while the centre of power shifted from Mithilā to Vaisāli, the modern Basārh in the adjoining district of Muzaffarpur. The Vrijjians included eight confederate clans, of whom the Lichchhavis were the most important. The growing power of the latter brought them into collision with the kingdom of Magadha, the limits of which roughly corresponded at this time with the modern districts of Patna and Gayā. Bimbisāra (*cir.* 519 B. C.), the founder of the Magadhan imperial power, had strengthened his position by matrimonial alliances with the more powerful of the neighbouring States, taking one consort from the royal family of Kosala, a kingdom to the west of Tirhut, and another from the influential Lichchhavi clan at Vaisāli. His murder by his son Ajātasatru involved the parricide in war with the aged king of Kosala, whose sister, the queen of the murdered Bimbisāra, is said to have died of grief. This war ending with the complete victory of Ajātasatru, his ambition next induced him to undertake the conquest of the country now known as Tirhut, in which the Lichchhavi clan occupied a prominent position (*cir.* 490 B.C.). The invasion was successful; the Lichchhavi capital, Vaisāli, was occupied; and Ajātasatru became master of Tirhut. It is probable* that the invader carried his victorious arms to their natural limit, the foot of the mountains, and that from this time the whole country between the Ganges and the Himālayas became subject, more or less directly, to the suzerainty of Magadha. From this time too dates the foundation of Pātaliputra (Patna), as the victor erected a fortress at the village of Pātali on the Ganges to curb his Lichchhavi opponents, and the foundations of a city nestling under the shelter of the fortress were laid by his grandson Udaya (*cir.* 434 B.C.).

---

* Early History of India by V. A. Smith.

Darbhangā must have shared in the religious ferment which so deeply stirred the hearts of the dwellers in the Gangetic valley in the sixth century B.C. North of the Ganges this movement seems to have centred at Vaisāli, which was thrice visited by Buddha, and was also the home of Mahāvīra Vardhamāna, the founder or rather developer of Jainism. The scene of Mahāvīra's ministry was laid in North and South Bihār; and here in a long wandering life of 42 years, he succeeded in gathering a considerable following of monks, known as the Nigranthas or men who discarded all social ties, who after Mahāvīra's death (*cir.* 490 B.C.) spread over the whole of India and became known as the Jains. About the same time, Buddha was engaged in his ministry in the same tract of country, and it seems beyond doubt that when he was spreading his doctrines at the neighbouring city of Vaisāli, the people of Darbhangā must have been among those who welcomed his teaching. It is at least clear from the account of the second great Buddhist Council that by the fourth century B.C. Tirhut had become a centre of Buddhism. This Council, it is said, was rendered necessary by the laxity of the Vrijjian monks, who asserted the legality of certain relaxations of the rule of discipline. The result was that the Buddhist community became split up into two contending parties, the strict and the lax, and the convocation of a Council was necessary for the restoration of order. The Council was held at Vaisāli, but the local monks failed to convince the assembled brotherhood of the validity of their claims, and all the relaxations were finally prohibited. The fact, however, that the Council was held at Vaisāli, and that the contentions of the local monks could cause such a schism in the Buddhist Church, sufficiently shows what a stronghold of Buddhism Tirhut was at this time. **Buddhism and Jainism.**

For several centuries after this we have no record of the history of Tirhut. It continued to form part of the territory of the Vrijjians and Lichchhavis, who acknowledged the suzerainty of the Maurya kings; but while the western portion, lying on the royal road from Pātaliputra to Nepāl, was enriched by the imperial monuments of Asoka, the eastern portion possesses no such remains. The capital, Vaisāli, seems to have been overshadowed by Pātaliputra, the seat of the Maurya kings; and during the reign of the Imperial Guptas, it probably formed the head-quarters of Tīrabhukti (Tirhut), one of the districts of their empire. Recent excavations* at Basārh have, however, disclosed remains which throw some light on its history during the fourth and fifth century A.D. Seals of officials have been found. which very probably were attached to letters addressed by imperial **Vaisāli.**

* Report Arch. Surv. Ind., 1903-04.

officers to the governors or chiefs of that district residing at Vaisāli, among whom are certain officers who are distinctly defined as being in charge of Tīrabhukti; and the great number of other seals, attached to letters sent by merchants and bankers, point to the large commercial transactions that were conducted in those days between the chiefs and important traders from Patna and other cities. Indeed, the presence of seals of mercantile and banking guilds lends some plausibility to the suggestion that something like a Chamber of Commerce had been established even at that early date in Upper India; while the titles of the chiefs or governors, such as the commander of the military forces, the head of the police, the chief chamberlain, the chief of the war-office treasury, the governor of the city and the chief of the Prince's ministers show a well-organized and advanced form of government.

Chinese pilgrims.

It is not until the seventh century A.D. that we find a detailed account of the country in the records of Hiuen Tsiang's travels, from which it would appear that Tirhut comprised the territory of Vaisāli to the south and the Vrijjian kingdom to the north. Hiuen Tsiang, who visited the country about 635 A.D., describes the kingdom of Vaisāli as about 1,000 miles (5,000 *li*) in circuit. The soil was rich and fertile, and flowers and fruit were produced in abundance, the mango and plantain being especially plentiful. The climate was agreeable and temperate, the manners of the people were pure and honest, and they loved religion and highly esteemed learning. To the north-east of Vaisāli was the Vrijjian kingdom, some 800 miles (4,000 *li*) in circuit, stretching away from east to west in a long narrow strip. The capital Chansuna (Janakpur?) was in ruins, and the inhabitants, who were of a quick and hasty temper, were mostly heretics who did not follow the law of Buddha. Hiuen Tsiang's account of Vaisāli shows that even there Buddhism was on the decline. Heretics and believers were living together, and he found the remains of hundreds of Buddhist monasteries; but only three or four were inhabited, and those sheltered but a few monks. The Jains were numerous, as might be expected in the birth-place of their religion; and Brahmanical Hindus worshipped at more than a score of shrines.* There is, however, proof that Buddhists

* In describing the seals of the fourth and fifth centuries, recently found at Basārh, Dr. Bloch remarks—"Turning to the emblems on the seals, the first thing that strikes one is the total absence of any symbol of Buddhism. * * The evidence of the emblems on the seals, so far as they have any connection with religious worship, together with the names occurring in the inscriptions and the seals bearing benedictory formulas, rather led me to conclude that most of the persons to whom the seals belonged were followers of the Brahminical creed or Jainas, or both." Report Arch. Surv. Ind., 1903-04.

remained there, probably until the conquest of the country by the Muhammadans, in the Buddhist images recently found among its ruins, which belong to the end of Buddhist history in India.

MEDIÆVAL PERIOD.

From the time of Hiuen Tsiang there is practically no historical information regarding North Bihār until the ascendancy of the Pāla dynasty (800—1200 A.D.). At the time of Hiuen Tsiang's visit it must have formed part of the dominions of Harsha, the powerful ruler of Northern India between 606 and 648 A.D., who exercised a certain amount of control as suzerain over the whole of Bengal as far east as Assam and possessed full sovereign power over Western and Central Bengal. But after his death the local chieftains no doubt asserted their independence, and Tirhut passed under the rule of petty potentates. Early in the ninth century Gopāla, the founder of the Pāla dynasty, became ruler of Bengal, and towards the close of his life (*cir.* 850 A.D.) he extended his power westward over Bihār. In the 11th century the country was wrested from the Pālas by the ambitious kings of Chedi, a tract corresponding to the present Central Provinces; and in 1019 A. D., it acknowledged the sovereignty of Gangeyadeva, who aimed at attaining paramount power in Northern India. The end of that century witnessed the rise of the power of the Sena Kings, who not only wrested their eastern provinces from the Pālas, but also appear to have carried their arms northwards to North Bihār. Mithilā formed the north-western province under the Sena rule, which in this part of Bihār is still commemorated by the use of the Lakshmana Sena era, the first current year (1119-20 A.D.) of which was apparently the date of either the accession or the coronation of Lakshmana Sena, the last great king of the Sena dynasty.

MUHAMMADAN PERIOD.

In the beginning of the 13th century the tide of Muhammadan conquest swept over Bihār, but it does not appear to have reached far north of the Ganges; for it is not till the time of Ghiās-ud-dīn Iwaz, the Muhammadan Governor of Bengal between 1211 and 1226, that we learn that the banner of Islām was triumphantly carried into the territories of the Rājā of Tirhut, which had never before been subdued, and the Rājā compelled to pay tribute. This appears, however, to have been rather a successful invasion than an effectual conquest of the whole country, as a local dynasty of Hindu kings was established about this time at Simrāon in the north-east corner of the Champāran district, and these kings succeeded in maintaining their rule over Tirhut for over a century, until the invasion of Tughlak Shāh in 1323 finally put an end to their independence.

The Simrāon dynasty.

Of the earlier kings of this dynasty we have only traditional accounts. Its founder was one Nāna or Nānyupa Deva, who is said to have established himself at Simrāon, and to have eventually subdued the whole of Mithilā and to have overcome the king of Nepāl. Tradition relates that one of his sons reigned in Nepāl, and the other, Gangā Deva, in Mithilā. Here he is credited with having introduced the system of fiscal divisions or *parganas* for the purposes of revenue administration; while a *chaudhri* or headman was appointed in each *pargana* to collect the revenue, and a *panchāyat* was chosen to settle all disputes. Local legend states that this king had a citadel at Laherā Rājā, a village in the Baherā thāna, and that the two large tanks of Gangā Sāgar near the Darbhangā railway station, and another at Andhrā Thārhi in the north-east of Madhubanī, were excavated during his reign. At the latter place he is said to have built a fort, and the villagers of Bairiyā, some 2 miles to the east, claim to be the descendants of his Prime Minister, Srīdhar Kāyasth. Gangā Deva was succeeded by his son Narsingh Deva, who is said to have had a quarrel with his kinsman, the king of Nepāl, the upshot of which was that Mithilā and Nepāl were separated never to be united again. Rām Singh Deva, who succeeded his father on the throne, was a pious devotee and a firm patron of sacred literature. Under his auspices, several well-known commentaries on the Vedas were compiled; rules were framed for the guidance of Hindus in their religious and social observances; and an officer was appointed in each village to adjudicate upon all questions arising from the working of these new canons of conduct. Various reforms in the system of internal administration are also attributed to this king. In every village a police officer was appointed whose duty it was to make a daily report of all occurrences worthy of note to the *chaudhri* or head revenue-collector of the *pargana*; the latter being assigned, in return for his services, a certain quantity of land, the produce of which was appropriated by him and his heirs in office. To the same period too is assigned the rise of the system of *paṭwārīs* or village accountants, who were, it is said, paid at the rate of Rs. 10 a month from the village funds.

On the death of Rām Singh Deva, his son, Sakti Singh, ascended the throne, but his despotism appears to have offended the nobles, and one of his ministers established a council of seven elders as a check upon the autocratic power of the king. His son Hara Singh Deva, the last of the line, is said to have excavated the Harahi and other tanks in this district, and to have prepared records of the caste subdivisions of the Brāhmans and Karna Kāyasths within his dominions. Fortunately, with this king we are on more certain ground than that of tradition. In 1323 the Emperor Tughlak Shāh led his victorious forces into Tirhut on his march back from the defeat of Bahādur Shāh, the rebellious Governor of Bengal. The fort of Hara Singh was taken, and the king himself fled northwards, and in his turn conquered the valley of Nepāl, though he does not seem to have maintained any effectual authority over it. As Professor Bendall* says:—"Until more evidence is forthcoming, it seems safer to regard Hara Singh and his ancestors who reigned in Tirhut, Simrāon and also possibly other parts of the Nepāl *Tarai* as at most titular kings of Nepāl, even if they really claimed sovereignty over the valley of Nepāl at all."

With the flight of Hara Singh, Tirhut became a dependency of the Empire of Delhi, and Tughlak Shāh placed it under Kāmeswara Thākur, the founder of the Thākur dynasty, which continued to rule over Tirhut till early in the 16th century. Here, as elsewhere, the Muhammadan conquest passed over the land without sweeping away all the ancient land marks. In some places the Muhammadans allowed the Hindu chiefs to remain undisturbed in their possessions, and in others they appointed new men as ruling chiefs; but in either case they did not interfere with the internal administration so long as their tribute was paid. The Hindu rulers of Tirhut were therefore practically independent, so long as they acknowledged their submission to the Muhammadans by the payment of an annual tribute. Their tenure of power, however, depended solely on the pleasure of their Muhammadan over-lords. The first of the line, Kāmeswara, was deposed by Firoz Shāh (1353), who gave the throne to his younger son Bhogiswara, who was his own personal friend. Kīrtti Singh, the second in descent from Bhogiswara, was also a younger son, who similarly obtained the principality as a personal favour from the Emperor as a result of a visit to Delhi. The most famous of the whole line, Siva Singh, was not so complaisant. He rebelled, and succeeded, in 1402 A. D., in establishing his independence; but three years afterwards he was conquered by the Musalmāns and carried off to Delhi, while his wife, Lakshimā Thākurānī, accompanied by the poet Vidyāpati, took refuge in Banaulī close to Janakpur in Nepāl, and there committed *satī* when no news of her husband had been received for 12 years. The memory of Siva Singh is still preserved among the people, who point to a tank he built at Rājokhari and quote the proverb:—"The tank at Rājokhari is indeed a tank, and all others are puddles. King Siva Singh was indeed a king and all others are princelings." But his chief claim to fame is

Thākur dynasty.

*See History of Nepal and surrounding Kingdoms, by Professor C. Bendall, J. A. S. B., Vol. LXXII, Part I (1903).

that he was a royal patron of learning. Not only was his wife, Lakshimā, one of the few learned women of India, but his court was frequented by poets and scholars, of whom Vidyāpati was at once the most famous and the most faithful. In this respect, Siva Singh was true to the traditions of his house. Like the Senas, who are said to have devoted their efforts to collecting troops of poems rather than to marshalling armies of soldiers, these Brāhman princes were noted for their encouragement of learning and the fine arts. Their courts were said to be the asylum of Sanskrit *belles lettres* and philosophy; and they lived immersed in the study of sacred books and poems.* One of the oldest known documents † in Darbhangā is a grant of Siva Singh's dated 1400 A.D., conveying the village of Bisapi (Bisfi in the Benīpatī thāna) to "the excellent poet Vidyāpati." His successors continued to rule as dependent princes over the northern part of Tirhut till 1532, when it was finally placed in the direct charge of Muhammadan Governors.

Muhammadan Governors.

Though the northern part of Tirhut was ruled by its native princes until this late period, the south was under the direct control of Muhammadan Governors, and here the Muhammadan supremacy was far more pronounced. In the neighbourhood of Hājīpur their ascendancy was won and maintained through war and convulsion. Hājīpur, situated at the confluence of the Gandak with the Ganges, was a position of great strategical importance; it was long the head-quarters of the Governors of the Bengal kings, and was the scene of more than one rebellion; and the result is that the Musalmān rule has left many more traces in the south than in the north of Tirhut. The city was founded by and named after Hājī Ilyās, king of Bengal (1345-58), who invaded and ravaged Tirhut, harried the frontier between the Bengal kingdom and the Delhi Empire, and apparently built this fortress to consolidate his conquests. To punish him and check his growing power, Firoz Shāh invaded Tirhut in 1353, following hard after Hājī Ilyās, who retreated to Pandua. The expedition was short and decisive, and after the entire defeat of Hājī Ilyās, the Emperor returned to Delhi after appointing officers to uphold his authority in Tirhut.

Shortly after this, North Bihār appears to have been annexed by the kings of Jaunpur (*cir.* 1397), and it remained subject to them for 100 years. They were then deprived of this outlying portion of their dominions by the Emperor Sikandar Lodī, who

* *See* Vidyāpati and his Contemporaries, and on some mediæval kings of Mithilā, by Dr. Grierson, Indian Antiquary, Vol. XIV, 1885, and Vol. XXVIII, 1899.

† The authenticity of this document is disputed, *see* J. A. S. B. Vol. LXVIII (1899), Part I, p. 96.

conquered Jaunpur, and then in 1499 advanced against Husain Shāh, king of Bengal. A treaty was concluded at Bārh, in which it was agreed that the Emperor should retain Bihār, Tirhut and *sarkār* Sāran, on condition that he did not invade Bengal; and Sikandar Lodī then swept down upon Tirhut. Unable to face the Imperial forces, the Rājā of Tirhut* advanced to meet him, made his submission, and was allowed to make terms on the payment of a fine amounting to several lakhs of rupees. The treaty between the Emperor and the Bengal king was not long observed, for in the early part of the 16th century Nasrat Shāh (1518-32) invaded Tirhut, put its Rājā to death and appointed his son-in-law, Alā-ud-dīn, to be its Governor. He then marched against Hājīpur, and having subdued the tract of country called after it, placed it in charge of another son-in-law named Makhdum Alim. Shortly afterwards (1538) the latter rose in revolt against his brother-in-law Mahmūd Shāh and made common cause with the Pathān adventurer Sher Shāh, who was at this time beginning the struggle which finally secured for him the throne of Delhi.

Mahmūd Shāh was the last independent king of Bengal, and after his fall North Bihār formed part of the Delhi empire. The allegiance of the local chieftains was however very uncertain, and they were practically independent. The country had not been fully subjugated, and it was filled with Afghān settlers, whose numbers were swelled by the accession of the Pathāns who had refused to join the service of the Mughals. Accordingly, when Dāūd Khān was raised to the throne of Bengal by the rebellious Pathān chiefs and broke out in open revolt in 1574, he found a number of trained soldiers ready to follow him in the strip of country north of the Ganges. Akbar sent orders to Khān Khānān to crush the rebel, and directed the Mughal chiefs of the neighbourhood and those Pathāns who continued loyal to the Mughals to assist him. It was not, however, until Akbar came in person to direct the operations that the fort of Hājīpur was taken and the rebels' power was broken. With the fall of Patna shortly afterwards, Bihār was lost to Dāūd Khān. It was thenceforward placed under a separate Mughal Governor, and Tirhut was included in the *Sūbah* or Province thus formed. Those chieftains who had assisted in establishing the Imperial authority were granted lands and *jāgīrs* in the Hājīpur *sarkār*, which included the southern portion of Darbhangā, and here they settled with their followers. Subsequently, in 1579, when they raised the standard of revolt, Azīm Khān, the successor of Akbar's great finance

* Elliot's History of India, Vol. V, p. 96.

minister Todar Mal, who was deputed to quell the rebellion, appears to have bought off the chiefs by confirming them in possession of the lands they had hitherto enjoyed and by granting fresh *jāgīrs*. In this way, a large number of petty Musalmān chiefs, with their followers, were permanently settled in the south of Muzaffarpur and Darbhangā.

The history of Darbhangā is henceforward merged in that of the Province of which it formed a part, and it does not come into any prominence until the troubled times which accompanied the disruption of the Mughal Empire. Then the Afghān chiefs settled there began to intrigue with the Marāthās, who were raiding Bengal, and when Mustafā Khān, the discontented general of Alī Vardi Khān, rebelled (1744), they sided with him. On their return to Darbhangā, they opened negotiations with Zain-ud-dīn, the son-in-law of Alī Vardi Khān, who was at this time the *Sūbahdār* or Governor of Bihār. They marched to Paṭna, where the Governor received them in *darbār* ; but towards its conclusion they treacherously assassinated him, and then proceeded to commit various savage atrocities: in the words of the author of the Riyazu-s-Salātin "they sacked the city and its suburbs, looted treasures, dishonoured women and children, and desolated a whole world." A vast number of Afghāns flocked to join them, till their army numbered about 40,000 horse and as many foot, besides a large train of artillery. But their triumph was short-lived, as Alī Vardi Khān marched up with a large army ; and in spite of the fact that they had been joined by the Marāthās, signally defeated them near Bārh. A number of the Afghān generals were killed, and their wives and daughters captured, but Alī Vardi Khān treated them honourably, allowed them to depart for Darbhangā, and even gave them property in the district for their maintenance. The revolt was effectually crushed, and the Afghāns of Darbhangā, with their leaders slain and their Marāthā allies in full retreat, did not again take up arms.

EARLY ENGLISH ADMINISTRATION. Darbhangā passed with the rest of Bihār under British rule in 1764, when the decisive battle of Buxar finally made the British masters of the Lower Provinces of Bnegal. Since that time the history of the district has been confined to the establishment of order, the progress of settled government, the growth of cultivation, and the general advance of civilization. In the yearly days of British administration, a great part of the country was in a terrible state of lawlessness. The trade routes were infested by robbers, who preyed on the surrounding tenantry, crimes went unpunished, and even high officials were molested. The correspondence on the first 20 years of British rule presents an extraordinary picture of the lawless state of the country, the trade routes blocked by gangs of freebooting zamīndārs, large stretches of land lying waste owing to their depredations, the native revenue officers in league with and sharing the booty of these outlaws, black-mail extorted from the servants of the local officers who had occasion to pass up the river, and even the Chief Magistrate himself fired on in order that he might be intimidated into paying an illegal toll; his attempts to arrest the offenders set at naught, and their apprehension not effected until a year later, and then only through the agency of military force. To the south it was reported that the zamīndārs had set authority in defiance and assumed an independence from any legal subordination, that the situation of their estates among the jungles and on the banks of the Bāghmati enabled them to keep a force which no person in those parts could repel, and that they managed to maintain it by means of depredations on both sides of the river. To the north matters were no better, as the zamīndārs on the borders of Nepāl, secure in their jungle fastnesses, mocked at the authorities. "They are all to a man," writes the Judge in 1782, "villains and tyrants, and many of them have long been in a state of petty warfare with Government;" and about ten years later the Collector describes them as "almost savages, who never occupied themselves except in hunting." Many of the peasants, unable to bear the constant oppression of these petty tyrants, abandoned their houses and lands, and fled to Nepāl, so that "in every village there were several empty houses and many villages were quite waste."

The difficulties of administration were still further increased by external trouble owing to the incursions of the Nepalese. Attempts to induce the Gurkhas to aid our officers in the suppression of frontier dacoities were fruitless, and all remonstrances against their aggressions were unavailing. The Collector of Tirhut reported that between 1787 and 1813 upwards of 200 villages had been seized by them upon one or other unjustifiable pretext, and in 1815 he was again obliged to report that the zamīndārs complained that in consequence of the incursions of the Nepalese, who had come down and burnt their villages and plundered their property, their ryots had in many instances deserted and they were unable to pay their revenue. In the war which ensued, Jaynagar, close to the Nepāl frontier, had to be occupied by British troops, but no fighting took place in the district.

Since the establishment of peace with the Nepalese, Darbhanga has enjoyed an unbroken tranquillity. During the

Mutiny there were great fears that mutineers from adjoining districts would break into Darbhangā. One body entered the *Tarai* north of Purnea, and it was thought that they might descend on Tirhut from Nepāl, but the rebels, failing to find the sympathy they expected, did not molest the district, and Darbhangā remained undisturbed.

SUMMARY. In concluding this sketch of the history of the district, the following summary* of its main features may be quoted :—" The history of Mithila does not centre round valiant feats of arms, but round courts engrossed in the luxurious enjoyment of literature and learning. But while Mithila's bid for fame does not rest on heroic deeds, it must be duly honoured as the home where the enlightened and learned might always find a generous patron, peace and safety. Before 1000 B. C. it was the great centre of Hindu learning ; from the sixth to the third century before Christ, Vaisali, a town within its borders, was a Buddhist stronghold, and in the fifth century, after the sun of Buddhism had set, Mithila again springs into notice as the home of Hindu enlightenment. When eventually the first flood of Musalman invasion, coming down the Ganges, did overspread Bihar, it subsided, leaving Mithila with Hindu kings still holding courts, where poetry and learning were alone honoured. In this country with principalities apparently undisturbed by internal troubles and heedless of external convulsion ; with courts devoted to learning and culture, where poets and philosophers lived in honour and affluence, our first impulse is to look for some traces of superior mental development in the mind of the people at large, at least for some grains of enlightenment fallen from the overflowing store of their masters. But the search is in vain, and it is a deplorable reflection on the Hindu social system that in Mithila, where it reached a high degree of development, its influences on the material and moral condition of the people at large were in the direction, not of amelioration, but of degradation. The priestly and intellectual aristocracy was so predominant that it set itself to suppress any attempt at social or mental emancipation outside its pale ; and it was able to attain its object so effectually that at the present day it is in this same ancient centre of learning that ignorant fanaticism is most rampant and religious susceptibilities most inflammable, that the state of the lower classes is most degraded and most depressed, and that the proportion of the illiterate is the greatest."

---

* Muzaffarpur Survey and Settlement Report, by C. J. Stevenson-Moore, I. C. S.

CHAPTER III.

THE PEOPLE.

UNTIL the year 1875 Darbhangā formed part of the old district of Tirhut, and no separate enumeration of its inhabitants was taken. A rough census of the population of Tirhut was, however, carried out in 1802, when it was estimated that the population numbered 2,000,000 souls. Seventy years afterwards, when the first regular census was taken, it was reported to be 4,384,706, but this first census was wanting in accuracy, and in all probability it was nearer 4,500,000. In other words, the increase of the population of Tirhut, *i.e.*, of the districts of Muzaffarpur and Darbhangā, was no less than 125 per cent. during these seventy years, or 1·78 per cent. per annum. GROWTH OF THE POPULATION.

According to statistics prepared for Darbhangā separately, it appears that the population of the district in 1872 was 2,136,898; and when the first census of Darbhangā as a separate district took place in 1881, it had risen to 2,630,496. A large portion of this increase was, however, due to greater accuracy in the enumeration. A careful census of the head-quarters subdivision in 1874 showed an apparent increase of 25 per cent. in the population during the space of 2½ years, one of which had been marked by general and severe distress; and a similar enumeration of the inhabitants of the Madhubanī subdivision in 1876 showed that there had been an increase of 20 per cent. during the 4 years which had elapsed since the census of 1872. The rate of increase disclosed is so extraordinary that it can only be explained as due to an inadequate enumeration in 1872. By the year 1891 the population had grown to 2,801,955, the increase during the decade being 6·5 per cent.; and the census of 1901 showed a further increase of 3·9 per cent., the total number of inhabitants being 2,912,611.

The decade preceding this census had not been a prosperous one. In 1891 the crops failed over a considerable area in the head-quarters and Madhubanī subdivisions, and relief operations were necessary. In 1895 the harvest was again a short one, and this was followed by the great crop failure of 1896. This affected the whole district except two of the three thānas in the Samāstipur CENSUS OF 1901.

subdivision: in the third thāna, Wārisnagar, the distress was less acute than in the rest of the district; and it was most severe in the western portion of the head-quarters and Madhubanī subdivisions. The number of persons requiring relief was greater than in any other district; but the recovery after the famine was at least equally rapid, and by the end of 1897-98 the people were almost restored to their normal condition. The result of the census was to show that, while there was only a slight increase (1·9 per cent.) in the Samāstipur subdivision, which was least affected by the famine, the growth was greatest (7·8 per cent.) in Madhubanī, where distress was most acute both in 1892 and 1897. The growth of population in the head-quarters subdivision was only 1·6 per cent., and here the Darbhangā thāna showed a slight decrease. This, however, was due to fortuitous circumstances, which, as explained below, caused an unusually large number of persons to be present in Darbhangā town at the time of the previous census; and if the urban area is excluded, the thāna shows a slight increase since 1891. The only other thāna in the district in which there was any decrease was Wārisnagar, where the population declined by 1·13 per cent.

Taking the variations of population since 1881, when the first census of Darbhangā as a separate district took place, one finds that the total increase has been just over 10 per cent., but more than three-fifths of this took place in the first decade. The Samāstipur subdivision lost 2·3 per cent. of its population in the ten years 1881-91, and though it recovered during the second decade some of the ground which it lost in the first, its present population is still less than it was in 1881. The growth in the head-quarters subdivision, which amounted to 8 per cent. in the first decade, has now fallen to 1·6 per cent.; and there has thus been a sudden arrest in the rise of population since 1891. In the Madhubanī subdivision the increase in the number of the inhabitants, which was very marked in 1881-91, continued, though it was considerably less rapid, falling from 12 to 7·85 per cent. Taking the district, then, as a whole, we find that the population of Samastipur is practically stationary, that the head-quarters subdivision is approaching a similar position, and that it is only in Madhubanī that any very large increase is to be found, and even here the rate of progress is appreciably less than formerly. Nor can these phenomena be ascribed to any special cause, such as emigration, famine, or plague. The movements of the population did not affect the results of the last census, and so far as permanent migration is concerned, the balance is in favour of the district. Plague may have affected the population of the

Dalsingh Sarai and Darbhangā thānas, but only slightly, and it did not extend to other parts before the census. In Madhubanī the reported death-rate in the preceding years had been higher than in any part of the district, and it had been most severely affected by the famine of 1897. In spite, however, of these unfavourable circumstances, Madhubanī showed the greatest addition to its numbers, an addition more than double that of the rest of the district. This is the great rice-growing subdivision, and there has apparently been some migration to it from the more crowded tracts in the south of the district; and the reason for the large increase appears to be that, while the other parts of Darbhangā have as dense a population as the land will support, there is still some room for expansion in Madhubanī.

GENERAL CHARACTERISTICS.

Although exceeded by the figures for a few individual districts, such as Howrah and Dacca, the portion of North Bihār which comprises the three districts of Sāran, Muzaffarpur and Darbhangā has a more teeming population than any other tract of equal size in Bengal or Eastern Bengal. Of these three districts Muzaffarpur is the most and Darbhangā the least densely populated, but the pressure on the soil even here is as great as 870 to the square mile. Lying between Muzaffarpur and Bhāgalpur, it partakes, in the west, of the character of the former, and in the east, of the character of the latter district. In only one of the western thānas does the density of population fall below 900, and in none of the eastern thānas does it reach 800 per square mile. The Samāstipur subdivision with 967 to the square mile is the most thickly crowded part of the district. It is eminently a rural tract, dependent almost entirely on agriculture; and the pressure of the agricultural population on the soil is already so great that further expansion under present conditions is neither to be expected nor desired. As, moreover, it includes about 40 square miles of *diāra*, which is practically uninhabited and mainly uncultivated, the pressure of population on the inhabited and productive area cannot be put at less than 1,000 per square mile. There is no very marked difference between the conditions of Madhubanī and the head-quarters subdivision so far as the strictly rural areas are concerned; and as the indications are that during the last decade the latter approached very near the limit of population which the soil is capable of supporting, it cannot be expected that Madhubanī will show any considerable further increase at the next census. These two subdivisions are mainly rice-producing tracts, and they cannot hope to support so large a population as Samāstipur, where the rich uplands produce more valuable crops. It may, therefore, be said of the district as a

Density of population.

whole that, under present conditions and in the absence of some economic revolution, there is little room for further increase of population.

Migration. Emigration as a remedy for over-population does not appear to have taken any real hold on the people as yet. The census of 1901 shows that out of every 10,000 persons born in the district only 382 are found elsewhere, and of these 255 are in contiguous districts. The proportion of emigrants is, in fact, lower than in any other district of North Bihār except Champāran and Purnea, where there is still much land available for cultivation. As regards immigration, 310 out of every 10,000 persons enumerated in the district were immigrants born elsewhere; and the volume of immigration is, therefore, nearly as large as that of emigration. So far as permanent movements are concerned, the balance is slightly in favour of Darbhangā, the number of emigrant females being nearly 4,000 less than that of those who have come in from other districts. Though the male emigrants number nearly 25,000 more than the male immigrants, the aggregate is only 58,443, and many of these are only temporary emigrants, who return to their homes after the labour season in other districts is over. The ebb and flow between Darbhangā and its neighbour is almost at a par; it loses to Bhāgalpur and gains in all other directions. The other emigrants go by preference to the neighbourhood of Calcutta or to Dacca and Eastern Bengal.

Sex. In common with other Bihār districts, Darbhangā has an excess of females over males, there being 106 females to every 100 males. Females preponderate in every thāna, and the preponderance has been steadily increasing since 1872. The causes appear to be obscure, but a partial clue may, perhaps, be found in the fact that the proportion of females to males is highest (over 110 per cent.) in Samāstipur and Dalsingh Sarai, and lowest (103 per cent.) in the Madhubanī subdivision. Pressure on the soil is greatest in the former area and least in the latter, and the relative preponderance of females probably indicates the extent to which temporary emigration is resorted to. The emigration to Bhāgalpur chiefly takes place from the northern and central parts of Darbhangā which adjoin it, and as it is largely of a permanent nature, the emigrants take their wives with them. But from the south of the district there is a good deal of temporary emigration to the eastern part of the Province during the labouring season, and in this case the emigrants usually leave their women-folk behind. Again, many of the marsh lands in the Gogrī thāna of Monghyr, which only produce a *rabi* crop and are submerged during the rains, are cultivated by men from the Dalsingh Sarai thāna, who only leave their homes for the cold weather and rarely take their families with them.

Towns and villages. Darbhangā is essentially a rural district. The great majority of the people are engaged in agricultural pursuits, and they have hitherto developed no tendency to flock into the towns. No new industrial centres of any importance have sprung up; and the rapid development of commerce and manufactures, which is so important a factor in the growth of an urban population, is unknown. Only 3 per cent. of the inhabitants live in urban areas, the remainder congregating in 3,233 villages. There are only four towns, viz., Darbhangā, Madhubanī, Iuserā and Samāstipur, with an aggregate population of 103,392, Darbhangā alone accounting for over two-thirds of the number. Of these four towns, Ruserā on the bank of the Gandak is slightly decadent, as the railway has diverted some of the traffic which was formerly borne by the river. Samāstipur and Madhubanī, on the other hand, have benefited by the railway, especially the former. Darbhangā at the last census showed a decrease of nearly 10 per cent., which was partly due to its unhealthiness in the preceding years and partly to special causes. The population returned in 1891 was abnormally large on account of the presence of some 5,000 Brāhmans, who had come to partake of a feast given by the Mahārājā of Darbhangā. In 1901 the fact that the date was an auspicious one for weddings caused the absence of many of the inhabitants; plague had also appeared, and some of the residents whose permanent homes were elsewhere had begun to move away; and the result was that the number of persons present in the town on the day of the census was unusually small.

LANGUAGE.* The vernacular current in the district is the dialect of Bihāri Hindī called Maithilī, *i.e.*, the language of Mithilā or the country bounded on the north by the Himālayas, on the south by the Ganges, on the west by the Gandak and on the east by the river Kosi. The head-quarters of this dialect, which is called Tirhutiyā by people in other parts of India, are in the north and centre of the district, where the Maithil Brāhmans muster in large numbers; in the south of Darbhangā it is spoken in a slighly corrupt form. It is spoken in its greatest purity by Maithil Brāhmans and the other high castes, while the lower castes have a habit of clipping their final syllables.

Most of the Muhammadans speak the Maithilī of their Hindu neighbours in a corrupt form mixed up with Arabic and Persian

* This sketch of the language of Darbhangā is taken from Dr. Grierson's Linguistic Survey of India, Vols. IV and V.

words. It is estimated that 337,000 out of the 338,667 Musalmāns in the district speak this corrupt Maithilī, or Jolāhā Bolī, as it is called from the name of the caste of Muhammadan weavers, who are numerous in Darbhangā. The upper classes, as well as the more educated Hindus, speak Hindustāni or Urdū, the number of speakers of this language being returned as about 4,000.

Written character. No less than three different alphabets are in use in the tract in which Maithilī is spoken. The Maithilī character proper, which is closely akin to the Bengali, is that used by the Maithil Brāhmans, and the character which is employed by all the other castes is the Kaithī. The Devanāgarī character is used by a few of the educated classes, and is understood and read by all persons who pretend to a liberal education; and, besides this, the Urdū character is employed by the better educated Muhammadans.

Literature. Maithilī is the only one of the Bihāri dialects which has a literary history. For centuries the Pandits of Mithilā have been famous for their learning, and more than one Sanskrit work of authority has been written by them. One of the few learned women of India whose name has come down to us, was Lakshimā Ṭhākurānī, who, according to tradition, lived in the middle of the 15th century A.D. Nor was the field of vernacular literature neglected by them. The earliest vernacular writer of whom we have any record was the celebrated Vidyāpati Thakkura, who graced the court of Mahārājā Siva Singh of Sugāonā, and who flourished about the same time. As a writer of Sanskrit works he was an author of considerable repute, but it is upon his dainty songs in the vernacular that his fame chiefly rests. He was the first of the old master-singers whose short religious poems, dealing principally with Rādhā and Krishna, exercised such an important influence on the religious history of Eastern India. His songs were adopted and enthusiastically recited by the celebrated Hindu reformer Chaitanya, who flourished at the beginning of the 16th century, and, through him, became the house-poetry of the Lower Provinces. Vidyāpati Thakkura or, as he is called in the vernacular, Bidyāpat Thākur, had many imitators in Mithilā itself, of whom we know nothing except the names of the most popular, and a few stray verses.

Amongst other writers in Maithilī may be mentioned Manbodh Jhā, who died about the year 1788 A.D. He composed a *Haribans*, or poetical life of Krishna, of which ten cantos are still extant, and enjoy great popularity. The drama has had several authors in Mithilā, where the local custom has been to write the body of a play in Sanskrit, but the songs in the vernacular. There has been a remarkable revival of Maithilī literature during the past few years, and at least one author deserving of special note has come to the front, Chandra Jhā, who has shown remarkable literary powers. He has written a *Mithilābhāshā Rāmāyana*, and a translation, with an edition of the original Sanskrit text, of the *Purusha-parīkshā* of Vidyāpati Thakkura, the perusal of which will well repay the student.

CHARACTER OF THE PEOPLE.

Maithilī, with its strangely complex system of verbal conjugation, has been described by Dr. Grierson as the dialect of a nationality which has carried conservatism to the excess of uncouthness. "Mithilā," he writes, "a country with an ancient history, traditions of which it retains to the present day, is a land under the domination of a sept of Brāhmans extraordinarily devoted to the mint, anise, and cummin of the law. For centuries it has been a tract too proud to admit other nationalities to intercourse on equal terms, and has passed through conquest after conquest, from the north, from the east, and from the west, without changing its ancestral peculiarities. The story goes that at the marriage of Rāmachandra,* the Brāhmans of Mithilā showed the same uncivilized pride which is the characteristic of their descendants of the 19th century. This Brahmanical domination has left ineffaceable marks upon the nature of the rest of the population. Mithilā or Tirhut is one of the most congested parts of India. Its inhabitants increase and multiply and impoverish the earth, nor will they seek other means of life than agriculture, or other lands on which to practise the one art with which they are acquainted."

There can be little doubt that the character of the people has been profoundly influenced by their geographical isolation. The river Gandak proved in the days of the Muhammadan invasion a curiously strong barrier: and while the countries to the west of the Gandak and south of the Ganges were constantly subjected to the turbulent influences that accompanied the rule of the Mughal dynasty, the country of Mithilā, the modern Tirhut, remained more or less at peace under Hindu kings. The results of this seclusion may be seen even in the present day. The people are more backward and less enterprising than those of Sāran. In the time of the Mutiny the rebels found fighting recruits in Sāran, but none in Muzaffarpur or Darbhangā; and at the present time the labourers of Sāran go far afield to seek labour, and are to be found even in the furthest corners of Assam, while the labourers of the other two districts leave their homes

---

* The curse pronounced by Rāmachandra on the Mithilā Brāhmans was:—
"Heroes at home, cowards in the battle-field, ever quarrelling among yourselves, and inordinately full of family pride, shall ye be in Mithilā."

in far fewer numbers. The people of Sāran are on the whole self-assertive, energetic and alert, while those of Darbhangā, living in a country which has for ages been secluded and far removed from the beaten paths of progress, are conservative, unenterprising and ignorant. A curious instance of this ignorance was seen a few years ago when some troops marched through the district, an event to which the inhabitants were unaccustomed. When the first relief went through, the villages were deserted by the women. The men came round in crowds to stare at the soldiers, and on one occasion, when the latter broke out into a cheer for some reason, the spectators fled across country in the greatest terror. After this, it became a regular practice for the soldiers to use this simple device to rid themselves of a staring throng when they were eating their food. Though the troops were a model of good behaviour, the movement of the second relief caused even greater excitement. They were Baluchis, and their advent was regarded with dread; and it took all the tact of the European residents and educated native gentlemen to assure the people that they had nothing to fear.

RELIGIONS. **Hindus and Muhammadans.** The great bulk of the inhabitants are Hindus, who with a total of 2,559,128 account for 87·86 per cent. of the population; and practically all the remainder are Muhammadans, who number 352,691 or 12·11 per cent. of the population. They muster most strongly in the Darbhangā thāna, where they form over a fifth of the population, and the proportion is lowest in Dalsingh Sarai, where they form little more than one-twentieth of the population. The Muhammadans conquered Tirhut, but they did not colonize it, and the tide of conquest passed over the land without shaking the firm roots which Hinduism had struck in the minds of the people. Islām has accordingly never held a strong position in Tirhut, and the country has always retained its pronounced Hinduism. Darbhangā, in particular, is the head-quarters of the Maithil Brāhmans, and Muhammadans form a comparatively unimportant section of the community.

Christians. The number of persons belonging to other religions is only 792, and of these 710 are Christians. There are several small Christian communities of different denominations, but no Missions of any importance; and the native Christians number only 296. There is a small Roman Catholic Mission at Samāstipur; a Methodist Mission and a Zanāna Bible and Medical Mission are also at work in the district.

RELIGIOUS MOVEMENTS. In recent years the only religious movements of any note occurred in the year 1893-94, when there was an ebullition of religious excitement among the Hindus, which found expression in this as in other Bihār districts in the anti-kine-killing agitation, the ploughmen's begging movement and the tree-daubing mystery.

Anti-kine-killing agitation.

The first movement appears to have been due to the activity of the Gorakshinī Sabhās or associations for the protection of cattle. These societies, the legitimate object of which is the care of diseased, aged, and otherwise useless cattle, started a crusade against the killing of kine, sent out emissaries to preach their doctrines, and collected subscriptions to further their objects. The relations between Hindus and Muhammadans soon became severely strained, and in various parts of Bihār the feeling aroused among the former manifested itself in the rescue of cattle from Muhammadans, and in objections to their slaughtering cattle and selling the meat. It was found that these institutions existed all over the district, and there was some fear that their preaching, at times ill-informed and malicious, might cause serious trouble by creating a spirit of antagonism among the Muhammadans. In spite, however, of a great deal of slumbering antipathy, the uneasy feeling which prevailed gradually quieted down, and no overt act of disturbance took place.

Ploughmen's begging movement.

The ploughmen's begging movement, or as it should more properly be called, the Mahādeo *pūjā*, was a curious exhibition of religious feeling which occurred soon afterwards. All ploughmen, the story goes, were obliged to give their cattle three days' rest and go round the neighbouring villages begging. With the proceeds three wheaten cakes were prepared—one for the ploughman himself, one for his cattle, while the third had to be buried under their stalls. This penance was performed by the people in consequence of a rumour that it had been imposed by the god Mahādeo to expiate the sin committed by the agricultural community in overworking their cattle. For some time the people continued to carry out, with scrupulous care, the orders which they supposed had been given them by their god. The remarkably elaborate nature of this penance gives reason, however, to suppose that it had been carefully thought out; and its inception and spread among the villagers has been attributed to the efforts of those interested in the Gorakshinī agitation to keep the movement afloat.

The tree-daubing mystery.

Tree-daubing was another widespread movement, the meaning of which afforded many grounds for speculation. By the most reliable reports it commenced about the latter end of February 1894 in the north-east corner of Bihār in the neighbourhood of the Janakpur shrine, which lies across the border in Nepāl. The movement consisted in marking trees with daubs of mud, in which were stuck hairs of different animals, buffaloes' hair and pigs' bristles predominating. It slowly spread through the Gangetic districts, eastwards into Bhāgalpur and Purnea and westwards through many of the districts of the United Provinces. In the South Gangetic districts, it was traced in several instances to wandering gangs of *sādhus*.

As an explanation of the movement, it was suggested at the time that the sign was intended as an advertisement of the shrine of Janakpur; and this view was accepted officially. Others, however, pointed out that it was suspicious that this movement should have followed the anti-kine-killing agitation, which contained a perceptible element of hostility to the administration held, and that it was part of a demonstration deliberately promoted against British rule. Others again believed that the marks were simply due to cattle rubbing against the trees.*

RELIGIOUS FESTIVALS.

The great majority of the people are agriculturists, dependent for their livelihood on seasonable rainfall and the fertility of the soil, and this dependence on the bounty of nature finds expression in their religious observances. In the month of Kārtik (October-November) it is usual to worship the cow, which provides milk for food, dung for manure, and oxen for the plough. In Māgh (January), after the oxen, the ploughmen and the plough have been given a six weeks' rest, the new year is ushered in by a quaint ceremony called *hal prabhā*. On the fifth day after the new moon the villagers take out the oxen and plough, and after mending the latter, the head of the household takes them to a piece of waste land, where he or the ploughman drives a furrow in a circle, which is passed over two and a half times. There the oxen are worshipped, being given fodder to eat and garlanded with flowers; and the ploughman and plough are then taken to the house, where the plough is worshipped by all the members of the household. It is held upright by the ploughman, the ploughshare resting on the ground; the iron tip of the share is covered with paddy; and flowers are sprinkled over it. The ceremony ends with the ploughman receiving all the paddy used to cover the plough. According to popular tradition, this ritual was initiated by Janaka, the ancient king of Mithilā, in time of drought; he bears the title *Sir-dhwaj*, or he whose ensign is the plough (*sir*); and the name of the day set aside for the ceremony, *Siri Panchamī*, on which the educated worship Swaraswatī, the goddess of learning, is said to be derived from the same word. Even the upper and middle classes offer oblations to the spirits of their ancestors in

* For a more detailed account of this movement and its meaning, *see* The tree-daubing of 1894, Calcutta Review, January 1898.

the belief that they will help their descendants by procuring timely rain and bumper crops in gratitude for the *pindas* offered to them. These ceremonies, called *parvans*, are four in number, and are celebrated on the day of a full or new moon. One called *Jar-pak*, which takes place in the month of Baisākh (April-May) when the *rabi* crop is reaped, consists of the offering of barley (*jar*) the produce of the spring harvest. The second is *Navadak* (the new rain), which takes place in the month of Asārh (June-July) with the breaking of the monsoon. The third, the *Shashtipak parvan*, is celebrated when the *bhadoi* crop is reaped in Bhādo (September-October); and the fourth, the *Navānna* (new crop), in Aghan at the time of the harvesting of the *aghani* crop on which the whole country-side depends.

Each portion of the agricultural year is marked by religious observances or superstitious practices which emphasize the anxiety of the people to enlist the powers of nature on their side. The anxiety with which the cultivator watches for the monsoon is intense, and should it fail, he endeavours to bring down the rain by appealing to the god of rain. It is believed by the vulgar that the cry of a frog is most readily heard by Indra. In a year of drought, therefore, the low-caste females of a village assemble in the evening, collect water from the pitchers of five houses, and seize a frog, which they put in a small earthen pot, together with water taken from the five pitchers. The pot, with the frog shut in it by an earthen cover, is put in the hollow wooden cup into which the lever used for crushing rice falls. Then the lever is raised with the foot and dropped on to the frog; this barbarously cruel performance being repeated until the frog croaks, which no doubt he does, unless killed with too great suddenness. Meanwhile, the women sing songs in a loud voice about the dearth of water. When the frog has been killed, its dead body is thrown on the roof of an equally credulous and superstitious neighbour, who in his turn abuses the thrower at the top of his voice, the belief being that this is pleasing to the god Indra (who is described in the Rig Veda as having eyes like a frog), and that he will be induced to send rain upon the earth.

The months of Bhādo and Asin (September and October) again are marked by many religious observances and ceremonies, because this is the most critical season of the year to the cultivator, when he must have rain. Towards the end of the former month the agriculturists have to observe the fast of *Ananta-brat* in gratitude for the ingathering of the *bhadoi* harvest and in the hope of further prosperity. During the first fortnight of Kuār or Asin, since it is on the rain of this period that a successful harvest of the *aghani* and moisture for the *rabi* depend, they devote much time to religious offerings and oblations to their deceased ancestors. This is followed by *Naurātra* or nine nights of abstinence from worldly enjoyments and worship of the goddess Durgā. When the *rabi* sowings have been completed and the *Naurātra* is over, there follows a day of universal rejoicing, when alms are given to every Brāhman who produces seedlings of barley, which here, as in other districts of Bihār and the Upper Provinces, forms one of the chief *rabi* crops.

During Kārtik (October-November), when the paddy harvest is taking ear, many devotional performances are observed, especially by the women and unmarried girls. They bathe before dawn and worship the sun as the producer of rain every morning until *Purnamāsi* or the period of the full moon, when large crowds of the people, both male and female, repair to bathe at the confluence of the Ganges and Gandak, and offer the holy Ganges water to Mahādeo or the Great God; it is then that the famous Sonpur Fair is held. Towards the end of this month the longest of all fasts is observed by the people, especially females, who before breaking it offer sweetmeats, vegetables and cow's milk to the Sun God. Even the pastime of the unmarried girls during this month is marked by a devotional spirit born of the mental suspense pending the arrival of the rice crop to maturity. All the children and young women go singing together to the fields for the feeding of what are called their *sama* and *chako*, clay images made to personify the agricultural gods, one representing the male and the other the female god. They do this every evening for the whole month of Kārtik, by the end of which the *aghani* should be fit for reaping; and on the 30th day they take the images to a neighbouring river or pond, and there submerge them; some even take the trouble to go to the river Ganges for the purpose. This is chiefly a female pastime, but the young children of either sex are allowed to take part. No adult male, however, is allowed to be present. Placing the images by turn in several plots (producing paddy, *maruā*, maize, etc.), the females make a circle round them and sing songs, dancing round and round till late in the night. One curious part of this observance is that, presumably to disarm scandal, an image of a *chuglā* or backbiter is set up, and the moustaches of the image are burnt, as a symbolical punishment for backbiting.

When the *rabi* crop is assured, the devotional attitude is abandoned, anxiety is at an end, and on the 1st Chait the people celebrate the *Holi* festival, breaking forth in unrestrained and hilarious enjoyment. The vitality which religious observances connected with agriculture still enjoy here arises no doubt from the almost entire dependence of agriculturists on the rainfall for their immunity from famine, whose grim visits have been too frequent to be easily forgotten.

Among Hindus, to plant a mango tree is considered a religious act, productive of spiritual benefit, and in this district nearly 88 per cent. of the population is composed of Hindus. The popular belief is that the rain-water falling from the leaves of a mango tree is converted into honey and received by the spirits of the ancestors of its planter, as well as by himself, when after death, he abides in *Swarga* or heaven. In fact, the approach of a mango grove to maturity is celebrated, by even the humblest cultivator, with all the show of a marriage ceremony. When the trees are sufficiently grown to give promise of fruit, the villagers repair to a place of worship erected in the grove, where a Brāhman officiates. The priest, after calling down the blessing of the gods, takes an earthen pitcher in which he places water, a few copper coins, and some *kasaili*. Over the top of the pitcher are placed mango leaves, and resting on them, a country lamp full of *ghī*, which is lighted. To make the symbolical ceremony complete, a bamboo basket containing a bride's belongings and dowry on a miniature scale is provided; a wooden post, called *yūpa*, is erected in the mango grove as a witness to the marriage; and the priest having finished his worship, vermilion, the emblem of completed marriage, is applied to the mango tree, as to a bride. Then a sacrificial fire is set alight, and the owner and his wife go round their grove; the former holding a mango leaf with a silver coin on it, over which a third person accompanying them from time to time sprinkles milk. This part of the rite is called *pradakshin*, the ordinary Sanskrit term for going round a sacrificial fire with the right side towards it. The grove is then dedicated to Narāyan, who is regarded as the bridegroom, and the ceremony ends with a feast to the priest and other Brāhmans. But the emoluments of the priest who officiates do not end here, for he is usually given money, *sajjadār*, *i.e.*, a bed with cushions, and one of the trees. It is thus not unusual to find an isolated tree in the middle of a mango grove in the possession of a Brāhman. The ryot willingly allows to the priest the enjoyment of the fruits, but objects to him cutting the tree down. The priest, on the other hand, owing to the difficulty of watching isolated trees, usually attempts to cut and sell the timber when the tree has attained a sufficient growth, thereby violating the religious faith of the planter.

Mock marriages of mango groves.

Mock marriages of tanks.

Similar ceremonies of marriage are performed in the case of newly excavated wells and tanks. The interesting feature in the marriage of tanks is the flight of the *dudh-pion* or the milk-drinker. A small quantity of milk is procured, and a Brāhman is induced by a money consideration to drink it. Having drunk it, he runs away to the distance of a mile, the people after him pelting him with clods of earth. He thus personates the *asūras* or enemies of the god; and the *jat* or wooden pole with a *trisul* or a three-pronged iron spear at its top, fixed in the centre of the tank, is an emblem of the weapon with which Mahādeo fought and overthrew them. It is now, however, popularly believed to be set up for the purpose of counteracting the evil eye.

PRINCIPAL CASTES.

The numbers and distribution of all the castes exceeding 25,000 persons are given in the Statistical Appendix. Among Hindus the most numerous castes are the Goālās or Ahīrs (383,653), Dosādhs (207,843), Brāhmans (197,967), Bābhans (154,345), Dhānuks (151,992), Koiris (144,673), Mallāhs (116,557) and Chamārs (105,739); while the Kewats, Khatwes, Kurmīs, Musahars, Rājputs, Tāntis and Telis all number between 50,000 and 100,000. Among the Muhammadans, Sheikhs (153,014) are most numerous, but Jolāhās (57,528), Dhuniās (39,884) and Kunjrās (39,397) are also well represented. Two small castes, the Deohars and Dhīmars, are peculiar to the district. The Deohars or Debhars of Darbhangā are said to form a separate caste. It is supposed that they may be an offshoot from the Goālās, but the two communities are now quite distinct. Their traditional occupation is that of inoculators, but at the present day they are largely employed as vaccinators. The Dhīmars have a tradition that they came originally from Benares in the retinue of a Rājā named Sibai Singh. They are probably an offshoot of some other caste, but it is difficult to ascertain which. Their name seems to be derived from Dhībara, a fisherman, which is also the origin of the word Tiyar. They are sometimes affiliated to the Kahārs, but beyond the fact that some are *pālki*-bearers, there seems to be no special connection with this caste. They believe the parching of grain to be their traditional occupation, but, as noted above, some now serve as *pālki*-bearers, and others work in lac, deal in fish, and serve as masons, menial servants and day-labourers.

Brāhmans.

The Brāhmans, though not so strong numerically as the Goālās and Dosādhs, are by far the most important caste in the district on account of their priestly influence. The great majority belong to the Maithil or Tirhutiyā sub-caste, of which there are 3 main divisions, based on a classification said to have been made by Rājā Hara Singh Deva, *viz.*, Srotriyā or Sote (high), Jog (middle) and Jaiwār (low). There is also a fourth division, Panjibaddh, comprising the offspring of intermarriages between the Srotriyā and the Jog classes, whose rank is intermediate between these two. These groups are to some extent hypergamous, *i.e.*, it is considered right that a man should bestow his daughter in his own or a higher group, but this is not compulsory, and if he is poor, he will often, for a sufficient consideration, give her to a man of a lower grade. It is simply a question of money, and the party which ranks higher receives money from the other side, whether it be that of the bride or of the bridegroom; the payment to a bridegroom is called *tilak*, and that for a bride *shulkā*. The amount varies according to the wealth and relative rank of the parties, and the minimum payment is said to be Rs. 35. Some of the poorer Brāhmans of higher rank will occasionally marry a number of wives of the lower grades for the sake of the money paid by their parents. This practice, which is said to be falling into disfavour, has given rise to a class called Bikauwās or vendors, who derive their name from the practice of selling themselves or their minor sons to girls belonging to lower groups. Some have as many as 40 or 50 wives, who live with their own parents and are visited at intervals by their husbands. Bikauwā Brāhmans who have married into lower groups are not received on equal terms by the members of their own class, but the women whom they marry consider themselves raised by the alliance. A succession of marriages into higher circles will gradually raise the position of a family, and, in the same way, a family gradually sinks in social estimation if all its matrimonial alliances are with persons of lower rank.

In addition to *gotras* or exogamous groups, the Maithil Brāhmans are distinguished according to their *mel* and *dih*. The *mel*, or origin, is the name of the village where the earliest known members of a *gotra* lived, while the *dihs* refer to the villages where their descendants subsequently settled. Thus the early progeny of Batsā Rishi lived, it is said, in a place called Karmahe, and this is the *mel* of all members of the Batsā *gotra*. Subsequent generations gradually spread abroad and settled in Majhaurā, Taraunī and other places, and the descendants of the residents in each place are distinguished accordingly as Karmahe Majhaurā, Karmahe Taraunī, etc.: these are called *dihs*. The classification of Maithil Brāhmans according to *mels* and *dihs*, like the other measures connected with their matrimonial arrangements, is ascribed to Rājā Hara Singh Deva.

It was effected in connection with the preparation of their pedigrees, and they attach much importance to it. It is a factor in estimating the relative rank of different members of the sub-caste, but it constitutes no bar to marriage. There is no objection to marrying a woman of the same *dih*, provided she does not belong to the same *gotra* or fall within the prohibited degrees of relationship.

Great importance is attached to purity of blood, and for this purpose the leading Maithil Brāhmans, with their pandits, their *pānjiārās* or genealogists, and their *ghataks* or marriage-brokers, meet at various places in Tirhut for the purpose of settling disputed questions of caste custom and of arranging marriages. They maintain one of the most extraordinary series of records in existence, called the *pānj*. It is composed of an immense number of palm-leaf manuscripts containing an entry for the birth and marriage of every pure Brāhman in Mithilā and going back for many hundred years—the *pānjiārās*, indeed, say for more than a thousand. These *pānjiārās* go on regular annual tours, entering the names of the Brāhmans born in each village during the last year; for no Brāhman can marry any woman who has not been entered in the *pānj*.

In arranging marriages the parties are assisted by *ghataks*, who carry in their memory the pedigrees of the Brāhmans of 10 or 12 villages, and who bring the parties together and help them to settle the amount of the consideration to be paid, for which service they receive a minimum fee of Rs. 2. When the parties have come to an agreement they go to the *pānjiārā* (Sanskrit *pānjikār*), who keeps the *pānj* or record of pedigrees; and after satisfying himself that the bride and bridegroom may lawfully be married, he grants them a written permission or *siddhānta patra*, for which he receives a fee of one rupee. *Pānjiārās* are said to have been first appointed by Rājā Hara Singh Deva, in consequence of one of his ministers having married in ignorance a lady within the prohibited degrees of relationship. The office is now hereditary, but, before practising, the candidate must obtain the permission of the Mahārājā of Darbhangā, who is the head of the Maithil Brāhmans, and who, after satisfying himself of his fitness, presents him with two loin cloths in token of his approval.

Marriages may be arranged at any time, but it is the fashion to meet for the purpose at certain regular assemblies held for the purpose towards the end of the *lagan* or marriage season. The largest of these gatherings is held at Saurāth, and extends over a week. Carpets are spread under the trees, and the Brāhmans assemble, gaily clad in crimson with flowing turbans. The occasion is one of unwonted rowdiness, and the crowd indulges in uproarious laughter, hooting and abusive shouts. When a marriage is decided on, the ceremony is at once performed at the house of the bride's father, whither the bridegroom proceeds, quite informally, accompanied only by a few of his near relations.

The Bābhans are the only other high caste with a strength of over 100,000. They are an influential agricultural caste, who form about 5 per cent. of the population; they are generally well-to-do, many of them being petty zamīndārs, but they are much addicted to litigation, and are notorious for their quarrelsome disposition. Like the Rājputs, they bear the names of Singh and Rai; and like the Brāhmans they are called Pānde, Misr, Thākur and Tewārī. They are commonly regarded as of mixed descent, and various legends are current as to the genesis of the caste. These traditions, however, are not recognized by the Bābhans themselves who claim to be true Brāhmans. According to their own account, they differ from Brāhmans only in having taken to agricultural pursuits and in having given up the principal functions of priestcraft, viz., teaching the Vedas, officiating as priests and receiving alms. They therefore call themselves Bhuinhār Brāhmans, and contend that many of their religious ceremonies are the same as among the Brāhmans, and that even at the present day Maithil Brāhmans who secede from their own community are admitted among them on condition that they give up priestly occupations. Bābhans.

There is no doubt however that, in the estimation of the Hindu public, they now constitute an entirely distinct caste. Their degradation from the status of Brāhmans must be very ancient, and it probably dates back to the period of the downfall of Buddhism. It has been pointed out that Bābhan is merely the Pālī form of Brāhman, and that the word is often found in Asoka's edicts. It has therefore been conjectured that those now known as Bābhans remained Buddhists after the Brāhmans around them had reverted to Hinduism, and so the Pālī name continued to be applied to them; while the synonym Bhuinhār or Bhumihāraka is explained as referring to their having seized the lands attached to the old Buddhist monasteries. In support of this theory, it may be mentioned that in this Province they are practically confined to the area covered by the ancient kingdom of Magadha, which long remained the centre of Buddhism.

The other castes require only a brief notice. The Goālās are by far the largest caste, and account for no less than one-eighth of Goālās.

the total population; they are most numerous in the north of the district. Their hereditary occupation is that of herdsmen, but with this they combine cultivation, and a large number have given up pastoral pursuits altogether and are only tillers of the soil. They are extremely quarrelsome and are noted *lāthiāls*; they are given to that exasperating form of theft which consists of petty thefts of crops from granaries and fields, and never lose an opportunity of grazing their cattle on a neighbour's crops; and they have an unenviable reputation as cattle-lifters.

Dosādhs. The Dosādhs, with an aggregate of 207,843, constitute a very numerous caste at the bottom of the social scale. They are one of the most useful castes in the district, owing to their value as agricultural labourers, but socially they form a very degraded community. Their religion is a sort of demon or fetish worship, and they sacrifice to appease the wrath of all the gods. The main features of their worship are sacrifices of pigs and libations of liquor, the ceremonies usually ending in a drunken orgy and a copious consumption of swine's flesh. They are despised by the Hindus generally, and appear to have traces of aboriginal descent. They support themselves by labour and cultivation; but many of them are notorious thieves, and, apparently on the principle of setting a thief to catch a thief, they have long all but monopolized the post of *chaukīdār* or village watchman in this district.

Dhānuks. Tradition relates that in prehistoric periods, when Tirhut was under a Brahmanic *régime*, the bulk of the labouring and industrial population were Dhānuks. Their name implies that they were originally bowmen, who obtained their livelihood as hunters; and there are some grounds for believing that they are of aboriginal descent. The common belief is that they are descended from a Chamār and a female Chandāl; but notwithstanding the degraded parentage assigned to them and the probability that they are really of non-Aryan descent, they now occupy a respectable position in the social scale. Personal service, palanquin-bearing and agriculture are their chief occupations; they are largely employed as *barāhils* or servants to zamīndārs, and are also engaged as agricultural labourers.

Koiris. The Koiris (144,673) are skilful and industrious cultivators, who are the best tenants to be found in the district. They are a purely agricultural caste, who also work as market gardeners and rear vegetables, potatoes, chillies and other crops requiring skilful treatment. They are the market gardeners of Bihār; and the production of opium in this district is almost entirely due to their labours.

The Mallāhs are a boating and fishing, as well as an agricultural, caste; but those engaged in the two former pursuits have a lower position in the social scale than the agricultural classes. Mallāhs.

The Chamār is one of the most indispensable of village servants, and in every hamlet or group of hamlets throughout the district will be found the houses of one or more Chamārs. He is a worker in leather, and supplies the village with their whipthongs and their shoes, and with leather fastenings for their carts: he is also one of the triumvirate of petty village officials, and with his coadjutors, the *chaukīdār* and the *gorait*, publishes by beat of drum all official proclamations. His wife, the Chamain, holds the important position of village midwife, and attends women of all castes during their confinement. In return for these services, the Chamār receives a small donation of grain from each ryot at the *aghani* harvest, and a present of Re. 1 or Rs. 2 annually from the zamīndār. The hides of all animals dying in the village are his traditional perquisite, and, in consequence of this, whenever a cow or other animal dies under suspicious circumstances, rightly or wrongly, he is almost invariably suspected of having caused its death. Chamārs.

CHAPTER IV.

PUBLIC HEALTH.

VITAL STATISTICS. A comparison of vital statistics for any lengthy periods is rendered impossible by the changes in the system of registering births and deaths which have taken place from time to time. In 1869 the duty of reporting deaths was imposed on the village *chaukīdārs*, and in 1876 the system was extended to births; but the returns received were so incomplete that they were soon discontinued and, except in the towns, deaths alone were registered until 1892, when the collection of statistics of births as well as of deaths was ordered, and the system now in vogue was introduced.

So far as they be accepted—and they are sufficiently accurate for the purpose of calculating the approximate growth of the population and of showing the relative healthiness or unhealthiness of different years—the returns submitted since that year show that conditions were generally unfavourable during the nine years ending in 1900. The mortality was exceptionally heavy in 1892, 1894, 1896, and 1900; in the first three years the deaths outnumbered the births, the death-rate in 1896 being 41·92 per mille, the highest ever recorded in the district; while plague broke out in 1900, and there was also a severe epidemic of cholera, causing 17,500 deaths. There was scarcity in the head-quarters and Madhubanī subdivision in 1891-92; there was famine in 1897; and in 1898 and 1899 there were floods in the south-west of the district. In spite, however, of these adverse conditions, the number of births reported exceeded the deaths by nearly 66,000. Since the year 1900, the population has been far more progressive, though there have been repeated visitations of plague. The death-rate has not exceeded 34·9 per mille, and there has been a very marked increase in births, the birth-rate averaging 43·35 per mille annually and rising in 1902 and 1904 to over 45 per mille. The number of births during the five years ending in 1905 consequently exceeded that of deaths by 153,000, the excess being 87,000 more than that recorded in the preceding nine years.

Taking the decade ending in 1905, the first half showed a mortality of 87,300 per annum, or 31 per mille, and in the second half the deaths rose to an average of 95,670, or 33 per mille; but the growth in the number of births was far greater, and the annual average increased from 106,500 in the first to 126,500 in the second quinquennium. Although, therefore, the mortality increased by 8,000 a year in the second period, the increase in the number of births (20,000) was more than twice as great.

PRINCIPAL DISEASES. Fever.

According to the returns submitted year by year, by far the greatest mortality is due to fever, but the ignorant *chaukīdār* responsible for the returns is far from being a medical expert. Drawn as he often is from the lowest dregs of the people, he can diagnose only a few well-known diseases like cholera and small-pox, and many others are indiscriminately classed under the general head of fever. It may, however, safely be assumed that when the mortality ascribed to fever is unusually high in any district, the greater part of the excess is due to malarial affections; and it is noticeable that since 1892, when the present system of mortality returns was instituted, the death-rate from fever has eight times exceeded that for the Province as a whole. During these years the mortality has never fallen below 20 per mille, but, on the other hand, it has never risen above 23·3 per mille, except in 1905, when it amounted to 26·6 per mille and in the three unhealthy years 1894—96. In the first of these three years it was nearly 30 per mille, and in the last it was no less than 31·15 per mille, the highest ever recorded in Darbhangā. Since that year the mortality from fever has remained fairly steady, and in the four years ending in 1904 it ranged only between 65,000 and 67,000 per annum, rising, however, to 77,600 in 1905. In that year only 400 villages escaped its ravages.

Regarding the different types of fever prevalent in the district, the Civil Surgeon, Captain M. H. Thornely, I.M.S., writes as follows:—Assuming that the term fever is to include all diseases in which a rise of the body temperature is a prominent or important symptom, the most prevalent fevers, approximately in order of frequency, are (1) malarial fevers; (2) fevers of short duration dependent on such conditions as the presence of worms in the intestine, exposure to heat, and local catarrhs and inflammations, such as coryza and bronchitis; and (3) epidemic fevers.

Though not so prevalent as in most other districts of Bengal, malaria is widespread, and accounts for the largest number of deaths. It is most severe and prevalent in the north of the district near the frontier of Nepāl, and in the south-east corner, regions which appear to be badly drained and mostly given up to rice cultivation. Most of the recognized types are found. It is only in the water-logged parts of the district just named that cases of extreme enlargement of the spleen with cachexia are at all common. As regards the diseases in which fever is a symptom of less importance, intestinal parasites, round worms, thread worms, and ankylostomdasis are extremely common in this district and are generally associated with pyrexia at some stage. Ankylostomdasis probably accounts for a large number of the deaths put down as due to fever. The epidemic fevers are small-pox, chicken-pox, measles, plague, enteric fever. Chicken-pox is not common, but occasionally becomes epidemic. The same remark applies to measles. Enteric fever is met with, but not frequently, and generally among Europeans only. An account of the epidemics of small-pox and plague is given below.

Cholera. After fever the greatest mortality is caused by cholera, which breaks out every year and occasionally spreads over the district with great virulence. During the first few years of the period for which reliable returns are available, there were widespread epidemics each alternate year, the death-rate in 1892, 1894 and 1896 being 4·6, 4·7 and 6·2 per mille, respectively, while it was very low in the alternate years. The district was then practically free from this scourge for three years, but since 1900 it has been an annual visitation, causing an average yearly mortality of over 9,000 in the five years ending in 1904, after which it fell to 3,000. It was most severe in 1900, when the returns showed 17,500 deaths, or 6·25 per mille, as due to its ravages.

Plague. For the last few years bubonic plague has regularly visited the district, the average annual mortality ranging from 1,000 to 4,000. Throughout these years, the disease has pursued a regular course, decreasing or disappearing entirely in the hot and rainy weather months, re-appearing after the rains, and reaching its climax in the cold weather.

Regarding these epidemics the Civil Surgeon writes as follows:—Plague has become established in the Darbhangā district since 1898. Apparently the first cases occurred in a village named Jalwāra in the Samāstipur subdivision in March of that year. During recent years it has shown itself in December or January, and has steadily increased during the months of

February, March, April and May. In June its virulence declines, and, as a rule, cases are not met with from July to November. The Samāstipur subdivision in the south of the district has been most severely affected, and the town of Darbhangā has also suffered greatly. An interesting fact, for which no satisfactory explanation has yet been adduced, is that the north and east of the district, comprising chiefly the Madhubanī subdivision, has up to now been practically immune from the disease. Imported cases have naturally occurred, but plague has never obtained a hold in these parts. Railways have spread, and communications have improved. The lack of these facilities for the importation of the disease cannot therefore be given as an explanation of the happy immunity these tracts have hitherto enjoyed. The reason most probably lies in the nature of the soil, the comparatively smaller density of the population of the villages, or the manner of construction of the houses.

The type is generally bubonic. The glands most commonly affected appear to be, in order of frequency, (1) the femoral and inguinal glands, (2) the axillary glands, and (3) the cervica glands. Pneumonic plague is very uncommon. The disease, as in other localities, is characterized by a low type of virulence at the commencement, by a very high type of virulence during the height of the epidemic, and again by a diminution of virulence as the epidemic ceases during the hot months and beginning of the rains.

Small-pox.

Small-pox also visits the district every year, but its ravages are not very severe, the death-rate due to it never having exceeded 0·30 per mille. It occasionally becomes epidemic among the unvaccinated, and is most frequent along the frontier where it is often imported from Nepāl.

Other diseases and infirmities.

The other diseases most frequently found are dysentery and diarrhœa, goitre, rheumatic affections, and skin, ear and eye diseases. Goitre is unusually prevalent; it is commonly attributed to the use of the water of the Little Gandak and Bāghmati, which come down from the Himālayas heavily laden with silt. The proportion of deaf-mutes is also greatest along the banks of these rivers, the affliction being most common in thāna Warisnagar which lies between them, and then in Samāstipur and Dalsingh Sarai, which lie between the Little Gandak and the Ganges. Altogether 117 males and 62 females out of every 100,000 of either sex were returned as deaf-mutes at the census of 1901; and these three thānas alone are responsible for two-fifths of the number, though they contain barely a quarter of the total population. Blindness is fairly common, and at the census of 1901 it was found that the proportion of blind persons among males was 101 and among females 81 out of every 100,000. The glare and dust accompanying a hot dry climate appear to predispose to cataract, and in the years 1896—1900 no less than 380 successful cataract operations were performed on males and 209 on females. Insanity is rarer than in other parts of the Patna Division, only 9 males and 5 females out of every 100,000 being returned as insane in 1901.

SANITATION.

Outside the municipalities, proper methods of sanitation are almost unknown, in spite of the measures taken by the District Board to clear rank undergrowth, fill up unhealthy hollows and sweep selected villages. The tendency of the people is towards aggregation, and instead of living in hamlets, each nestling within its own belt of trees, as in Bengal, they cluster in closely-packed villages, usually consisting of a main street with narrow and ill-ventilated side lanes. Sanitary conditions are extremely primitive, while the apathy of the people and the unwholesome habits to which they are rooted render the task of village sanitation on any appreciable scale most difficult. Village sites are generally clusters of thatch-roofed mud-walled houses with extremely insanitary surroundings. The drainage is left to its own care, and ventilation is unthought of. If windows exist, they are made, at any rate in the zanāna apartments, not more than a couple of feet from the top of the wall, and then securely latticed, their use for ventilation being thereby minimized as far as possible.

Here and there, a man who has got on in the world may move out of the village *bastī*, and build himself a brick-house elsewhere, but the average peasant, however prosperous he may be or however much his family may increase, will cling to the house of his father and his father's gods. To spend his savings on the improvement of his dwelling is the last thing that would occur to him: he regards his house only as a place to cook and sleep in, and for half the year he and his sons sleep in the open. The houses themselves are usually constructed of earth, dug out of a hole in the immediate vicinity, which in the rains becomes a regular cess-pool, covered with rank vegetation and giving out a most offensive smell. Even the wells from which the people obtain their drinking-water are shamefully neglected, and are often allowed to get into a disgustingly filthy condition. In many instances, they are surrounded by house-drains, a large portion of the contents of which must find its way into the wells by percolation, carrying with them the germs of cholera. In the municipalities there is an organized system of conservancy; night-soil and other refuse are removed; and steps are taken to protect the sources of water-supply. But none of them have a regular system for the supply of pure water or an adequate scheme of drainage, and, like the villages, they suffer from crowded and badly-aligned blocks of houses intersected by narrow lanes. Owing to the congestion of the population, the mortality from plague, dysentery and cholera are generally greater than in rural areas; in 1905 the death-rate in the municipalities was 47·03 per mille, and in the rest of the district 34·36 per mille, while the average for the preceding 5 years was 43·14 as compared with 32·05 in rural areas.

VACCINATION. Formerly there was a strong prejudice against the practice of vaccination, and inoculation was prevalent; but the people have learnt to appreciate the protection afforded by the former, and this prejudice has passed away. Altogether 84,000 persons, or 30 per 1,000 of the population, were successfully vaccinated in 1904-05; and in each of the preceding five years protection was afforded to an average of 30·3 per mille. These figures are higher than in any other district of the Patna Division, except Muzaffarpur. This result is all the more satisfactory, because it is easy for any people in the north of the district who object to vaccination to avoid the vaccination agents by quietly going to a neighbouring village across the frontier. Statistics of vaccination from 1892-93 to 1901-02 will be found in the Statistical Appendix.

MEDICAL INSTITUTIONS. The district contains six hospitals—the Dufferin and Rāj hospitals at Darbhangā, the Banwāri Lāl Hospital at Laheriā Sarai, male and female hospitals at Madhubanī, and a hospital at Samāstipur; and thirteen public dispensaries at Baherā, Benīpatī, Dalsingh Sarai, Khajaulī, Madhubanī, Madhepur, Malmal, Nandini, Narahiā, Narhan, Pūsa, Ruserā, Sakrī and Tājpur. Great activity has been displayed in recent years in opening new dispensaries, and the annual number of patients treated rose from 117,553 in 1894 to 130,438 in 1899 and to 301,536 in 1904; in other words, the attendance at dispensaries increased by 131 per cent. in the decade. The average daily attendance in the same three years was 754, 822 and 1,577, respectively. The annual expenditure on the dispensaries is about half a lakh of rupees, half of which is derived from private subscriptions and endowments, nearly the whole of the remainder being contributed by the District Board and the various municipalities. Statistics of the daily attendance, indoor and outdoor, during the years 1892—1901 will be found in the Statistical Appendix; and the following tables give a statement of the receipts and expenditure and of the principal diseases treated in 1905:—

Name of Dispensary.	Receipts.				Expenditure.	
	Government contribution.	District fund.	Municipal fund.	Subscriptions and other sources.	Establishment.	Medicines, diet, buildings, &c.
	Rs.	Rs.	Rs.	Rs.	Rs.	Rs.
Baherā Dispensary ...	15	753	...	...	574	194
Benīpatī ditto ...	15	644	...	...	564	95
Dalsingh Sarai ditto ...	15	...	...	1,534	755	794
Darbhangā Dufferin Hospital	15	...	...	7,302	3,115	4,202
Darbhangā Rāj Hospital ...	39	...	...	16,476	5,116	11,399
Khajaulī Dispensary	5	1,972	...	...	60	1,917
Laheriā Sarai B. L. Hospital	801	2,402	2,635	370	3,827	4,047
Madhepur Dispensary	15	742	...	...	573	184
Madhubanī Female Hospital	15	778	...	...	693	100
Madhubanī Male Hospital ...	88	600	2,250	868	2,655	1,192
Malmal Dispensary	15	647	...	2	567	97
Nandini ditto	15	...	...	3,000	828	2,187
Narahiā ditto	16	...	...	2,004	785	1,235
Narhan ditto	15	...	...	1,997	852	1,169
Pūsa ditto	421	...	...	...	375	46
Ruserā ditto	340	225	219	680	1,176	446
Sakrī ditto	15	680	...	...	545	150
Samāstipur Hospital	104	...	700	805	729	1,299
Tājpur Dispensary	15	277	...	422	574	140
Total ...	1,979	9,718	5,804	35,460	24,363	30,884

Name of Dispensary.	Diseases treated.					Number of operations.
	Worms, dysentery and other bowel diseases.	Fever.	Diseases of eye and special senses.	Injuries.	Other diseases.	
Baherā Dispensary ...	4,465	1,824	680	34	7,003	181
Benīpatī ditto	4,930	1,570	545	103	3,822	380
Dalsingh Sarai ditto ...	1,599	1,637	821	64	10,909	415
Darbhangā Dufferin Hospital	11,155	5,240	867	34	2,420	234
Darbhangā Rāj Hospital ..	38,276	20,380	3,572	487	28,431	1,355
Khajaulī Dispensary ...	121	132	37	21	311	40
Laheriā Sarai B. L. Hospital	33,640	8,819	2,561	747	17,521	46
Madhepur Dispensary	1,466	1,112	522	46	4,161	109
Madhubanī Female Hospital	4,448	1,720	824	17	4,704	200
Madhubanī Male Hospital ...	6,433	1,538	1,510	271	6,383	589
Malmal Dispensary ...	2,728	795	1,022	35	4,811	213
Nandini ditto ...	799	1,340	304	144	4,333	261
Narahiā ditto ...	1,149	580	1,271	74	4,676	672
Narhan ditto ...	2,350	3,058	782	122	7,933	312
Pūsa ditto ...	213	172	78	57	797	41
Ruserā ditto ...	1,286	1,502	971	131	6,105	562
Sakrī ditto ...	4,792	1,179	826	123	4,222	387
Samāstipur Hospital ...	1,622	689	636	212	8,318	382
Tājpur Dispensary	2,705	3,028	1,306	66	7,105	435
Total ...	124,227	56,315	19,125	2,788	133,965	7,714

CHAPTER V.

AGRICULTURE.

GENERAL CONDITIONS. Rainfall.

ALTHOUGH there is a certain amount of irrigation in the north of the district, the main sources of supply are tanks and streams, which are apt to fail in a dry season when they are most wanted. Artificial irrigation therefore plays a comparatively minor part in the agricultural economy of the district; and Darbhangā, like the rest of North Bihār, is dependent for its crops on the local rainfall. If the monsoon is up to its normal strength, and the rain is timely and well distributed, it admirably serves the purposes of the agricultural system generally practised. The main crop in the district is winter rice, which covers over three-fifths of the cropped area. This crop requires good rain at the end of May and in June, to facilitate the preparation of the ground and the growth of the seedlings. More rain is required in July and August for transplantation, and without a good fall during the *Hathiyā* asterism at the end of September and during the early part of October, the rice withers away and never comes to maturity. All the essentials for a good winter rice crop are thus secured in a year of normal rainfall, which also provides admirably for the crops of the *bhadoi* harvest, which are sown in May and June and reaped in August and September. A good rainfall at the end of monsoon further secures moisture for the crops of the next spring harvest, which are sown in October and November, and in a normal year are refreshed during the cold weather by seasonable showers. Unfortunately, the district does not always enjoy a normal rainfall, and whether this falls short of the average, or is badly distributed, full crops cannot be looked for; and in a district so dependent as Darbhangā on its winter rice, the most serious results naturally ensue from a failure or premature cessation of the monsoon. Although, therefore, Darbhangā is practically independent of irrigation in a year of normal rainfall and can raise full crops without it, the importance of seasonable and well distributed rain, particularly during the monsoon, cannot be overestimated.

Another very important factor affecting the agricultural prosperity of the district is its liability to floods. Provided that they come fairly early in the season and are not of great height or long duration, floods are productive of as much good as harm, as the cultivators have time to re-transplant if the crops are destroyed, and the land is enriched by the silt they leave. Very great damage however is caused by inundation in exceptional years, such as 1906, when floods of an unprecedented height and of long duration swept over a very extensive area, with the result that the *bhadoi* and *aghani* crops were almost entirely destroyed in the head-quarters subdivision, while they were very seriously damaged in other parts of the district, and that acute distress ensued throughout the flooded country, and in some tracts culminated in famine.

Tracts of fertility. The first of the three tracts mentioned in Chapter I, viz., the south-western portion of Darbhangā, comprising the Dalsingh Sarai and Samāstipur thānas, is the richest and most fertile area in the district and grows all the most valuable autumn (*bhadoi*) and spring (*rabi*) crops. The second tract, viz., the *doāb* between the Bāghmati and Little Gandak, is liable to inundation from the former river, and the main crop produced is winter rice, though good *rabi* crops are also raised in many parts from the lands enriched by the fertilizing silt deposited by floods. The third tract, as already stated, comprises the head-quarters and Madhubanī subdivisions, the south-eastern portion of which is in the rainy season a vast chain of lakes linked by the numerous streams flowing south. In this part the only crop of any importance is rice, which when not submerged by early floods is very prolific. In the Madhubanī subdivision, where the land is generally higher, the staple crop is winter rice; but the three western thānas and the south of the Phulparās thāna, contain stretches of high land suitable for the more valuable *rabi* crops.

Soils. The soils of the district may be divided into three kinds, *balsundrī*, a sandy loam; *bāngar* or *chikna*, a clayey soil with an admixture of sand; and *matiyāri*, a clay soil with little or no sand. The three main classes correspond roughly with the three physical divisions of the district. *Balsundrī* is the prevailing soil south of the Little Gandak in the thānas of Samāstipur and Dalsingh Sarai. The *doāb* between the Bāghmati and the Gandak, corresponding roughly with the Wārisnagar thāna, is chiefly composed of *bāngar*, while the prevailing soil in the rest of the district is *matiyāri*. But the distribution is not a strict one, for *matiyārī* soils are found in *chaurs* in Samāstipur, and *balsundrī* in the high lands of Benīpatī *Matiyāri* soil, being extremely retentive of moisture, is best suited for the growth of winter rice. *Bhadoi* crops, which cannot stand too much moisture, do best in *balsundrī* or sandy *bāngar*, and the more valuable *rabi* crops are almost exclusively grown on *balsundrī*. Rice also does well on low-lying *bāngar* lands, if the admixture of sand is not too strong to absorb the moisture. All over the district are found patches of *ūsar* land, which are unproductive owing to the salt efflorescence known as *reh*: it is generally believed that this can only be eradicated by inundation.

On the whole, however, the foregoing classification of soils is of little practical importance, and would not be readily understood except by the more intelligent cultivators. The only classification of land understood by the ordinary cultivator is that of *dhānhar* and *bhīth*, *dhānhar* meaning the low land on which rice (*dhān*) is grown, and *bhīth* the uplands growing cereals or crops of any kind other than paddy. This was the classification adopted in the records at the time of the settlement concluded in 1903.

There are no Government irrigation works, but an area of 171 square miles or 6½ per cent. of the cultivated area is irrigated from other sources. Most of this area lies in the north of the district, where irrigation is devoted to food crops, whereas in the south it is reserved for the special and highly cultivated non-food crops. Here extensive irrigation is not practicable, owing to the scanty sources of supply, nor it is greatly required for the crops mainly grown. There is a prejudice against well-irrigation, as it is believed that land once artificially irrigated must always be irrigated. This belief is probably to a certain extent well founded; for in the soils prevailing in this area, irrigation forms a crust below the surface, which impairs the fertility of the land, unless irrigation is continued every year. Hence irrigation is only practised on lands near wells, and these lands are reserved for the more valuable crops. Irrigation once begun must be continued, and though it may result in a good return through the production of more valuable crops than can be raised on unirrigated lands, it entails an amount of labour and expense from which the majority of cultivators are averse. Irrigation.

Artificial irrigation is most practised in the Madhubanī subdivision to the north of the district. Here the numerous streams and rivers which intersect the country are utilized for the supply of water to the winter rice crop, especially in the Khajaulī and Phulparās thānas; and in the Benīpatī thāna nearly one-third of the net cropped area is irrigated from a complete system of *pains* or artificial channels led off from the Kamlā river. These *pains*, for many of which the old beds of the Kamlā were utilized, were constructed during the famine of 1897 by the energy and foresight of Mr. King, the Sub-Manager of the Rohika Circle of the Darbhangā Rāj ; and they were the means of saving 30,000 acres of winter rice during the partial failure of the monsoon in 1901. In the area covered by these channels the outturn of winter rice in that year was 80 per cent. as against 30 per cent. on the west, 21 per cent. on the east and 19 per cent. on the south. Besides these sources of supply, tanks are very largely used all over the subdivision for the irrigation of the fields in their neighbourhood, either to expedite the transplantation of the winter rice seedlings or to prevent them from withering during a break in the rains.

In the head-quarters subdivision, the main sources of irrigation are the same as in Madhubanī, but the area irrigated is very much less. Most of the Madhubanī streams join the larger rivers before they reach this subdivision ; and the latter are too large to permit of their being used for irrigation by the simple and inexpensive means adopted by the Madhubanī cultivators. Tanks are also less numerous, and the ryots being less familiar with irrigation than their fellows in Madhubanī, make less use of such tanks as there are.

Little, therefore, can be done in the way of extending irrigation in this part of the district owing to the scanty sources of supply ; and this is the more unfortunate, because it is this area which, as a rule, suffers most severely in case of failure of the winter rice crop. In the north of the district many opportunities exist for the extension of irrigation on a comparatively large scale, but here a great difficulty is presented by the danger that the supply of water would be cut off in Nepāl at the time when it is most necessary. The question is further complicated by the fact that, in years of normal rainfall, irrigation is little practised or required. The water, which, in a year of drought, is priceless, is in an ordinary year a superfluity ; and in a dry year, when irrigation is most wanted, there is a scanty supply of water in the tanks and streams. All irrigation projects in Madhubanī, accordingly, suffer from the defect that they would be little used in ordinary years, while in dry years they would benefit a comparatively small area. Hence large irrigation works are probably financially impossible, and any extension of irrigation must be looked for, mainly in the direction of increasing the usefulness of the present sources of supply by some cheap and efficient system, such as that which has already proved so successful in Benīpatī. Well irrigation is practically useless for winter rice, owing to the small area commanded by each well, and no wells would stand in an area so liable to inundation. But tanks can already irrigate 45,000 acres in the subdivision, and rivers and channels nearly 50,000 acres. It would probably be possible to improve the manner in which these two sources of irrigation are used, at a cost which would be insignificant in comparison with the advantage to be gained, and this would go far to render a large part of the area secure against a failure of the monsoon.

PRINCIPAL CROPS. In Darbhangā, as in the rest of Bihār, the crops fall under one or other of the three main harvests, the *aghani*, *bhadoi* and *rabi*. The *aghani* is the winter crop, which is cut in the month of Aghan, and is composed mainly of winter rice. The *bhadoi* is the early or autumn crop, reaped in the month of Bhādo (August-September), consisting of 60 days' (*Sāthī*) rice, *maruā* (*Eleusine Coracana*), maize, indigo, millets and less important grains; while the *rabi* crop, which is so called because it is harvested in the spring (*rabi*), includes such cold weather crops as wheat, barley, oats, gram, pulses and poppy. Of the net cropped area 62 per cent. grows *aghani*, 47 per cent. grows *rabi*, and only in 27 per cent. are *bhadoi* crops raised. With regard to the proportion of the area under these different crops, it must be remembered that the chief *aghani* crops are sown before the *bhadoi* crops are harvested, and similarly the chief *rabi* crops are sown before the *aghani* is reaped. Hence the latter cannot usually be sown as a second crop to *bhadoi*; and where it is predominant, the *rabi*, grown on land which has previously yielded winter rice, consists mainly of the cheapest kind of grains, such as *khesāri* (*Lathyrus sativus*), a grain of inconsiderable value, or a thin catch-crop of gram or linseed.

The *aghani* predominates in the Madhubanī and head-quarters subdivision, occupying 72 per cent. of the net cropped area in the former and 66 per cent. in the latter subdivision; and consequently the area under *bhadoi* crops is comparatively small. This tract is, moreover, unsuitable for the latter crop on account of its liability to floods; and the proportion of the cropped area under *bhadoi* is only 22 per cent. in the former and 25 per cent. in the latter subdivision. It is highest (41 per cent.) in the Samāstipur subdivision, and as it is only *bhadoi* lands that yield a *rabi* crop of any importance, the proportion under *rabi* (46 per cent.) is approximately the same. In each of the other two subdivisions the *rabi* accounts for 47 per cent. of the net cropped area, but, as already explained, the percentage of land under winter rice is too high to leave room for any but the inferior *rabi* crops.

FOOD CROPS. Food crops occupy altogether 84·21 per cent. of the net cropped area. Rice alone is raised on 61·39 per cent., and covers both a larger gross area and a larger proportion of the cropped area than in any other district of the Patna Division north of the Ganges. After rice, *maruā* occupies the largest area (220,000 acres) and the two crops together are grown on nearly three-quarters to the net cropped area of the district. This predominance of rice and *maruā* leaves little room for the more valuable *bhadoi* and *rabi* food crops, such as wheat, barley and maize; and the proportion of land under these crops is smaller than in any other North Gangetic district in the Patna Division.

Rice.

Rice, covering three-fifths of the net cropped area, is the all-important crop in Darbhangā, and no less than 93 per cent. of the crop consists of winter rice. This is most prevalent in the north of the district, where the liability of the country to inundations renders it unsuitable for the growth of *bhadoi* or autumn rice, except on the higher lands. The latter is most extensively grown in Khajaulī thāna, where, as in the neighbouring thāna of Benīpatī, it is often destroyed by floods. The cultivators are, however, generally compensated for the loss of their *bhadoi* crops in this area by the excellent *aghani* crop which they obtain from the submerged lands. With so large an area under winter rice, less than one-tenth of which is irrigated, Darbhangā is a district peculiarly sensitive to variations in the monsoon rainfall; for seasonable and well-distributed rainfall throughout the whole course of the monsoon is essential to the prosperity of this crop. The *bhadoi* crops on the other hand, while requiring good rain in the early part of the monsoon, are reaped during September, and are consequently independent of the *Hathiyā* rains at the end of that month, the absence of which is fatal to the *aghani* harvest. This explains why the head-quarters and Madhubanī subdivisions are always liable to scarcity or famine in the event of a failure of the monsoon, while the greater part of Samāstipur is immune, for though rice is the most extensively grown crop in the latter subdivision, it covers less than a third of the net cropped area, as compared with the district average of three-fifths.

Methods of rice cultivation.

Rice is cultivated either by being sown broadcast or by means of transplantation. The former method is followed in the case of *bhadoi* rice. It is sown broadcast in June or July and is not transplanted; it is regarded as a 60 days' crop and is reaped in August or September. Transplantation is the commonest method, which is followed in the case of nearly all the *aghani* or winter rice. The seeds are first sown broadcast after the commencement of the rains in June or July on lands selected for seed nurseries, ich have previously been ploughed three or four times. After 4 or 6 weeks, when the young plants are about a foot high, they are generally transplanted, each plant being pulled out from the land, which is soft with standing water, and planted again in rows in flooded fields in which the soil has been puddled. The rice is then left to mature, with the aid of water, till towards the end of September. The water is then drained off and the fields are allowed to dry for 15 days and at the end of that time they are again flooded. It is this practice, known as *nigār*, which makes the rainfall which is expected at this time, or failing that, irrigation essential to a successful harvest. These late rains (the *Hathiyā*) are the most important in the year, for not only are they required to bring the winter crop to maturity, but also to provide moisture for the sowing of the *rabi* crops. Should no rain fall at this period, or if water cannot be procured from artificial sources, the plants will wither and become only fit for fodder; but if seasonable showers fall, or if the crop receives artificial irrigation, the rice comes to maturity in November or December.

Miscellaneous food-grains. After rice, the next largest area is that shewn as under miscellaneous food-grains. These include all cereals and pulses except *maruā*, maize, barley, wheat and gram, for which separate figures are given. Of the miscellaneous food-grains, the most widely grown is *khesāri* (*Lathyrus sativus*), which is usually sown as a catch-crop in *aghani* lands, at or just before the time of the *aghani* harvest. It is a cheap grain, and as a rule is eaten only by the poorer classes, though it is sometimes used with barley to make *chapātis*. *Arhar* (*Cajanus indicus*) is another important crop shewn under this head. It is usually sown with maize in May or June, but is not reaped till the following March. One main objection to the crop from the cultivator's point of view is that it occupies the land for so long a period, while the grain is not so valuable as that of other crops. On the other hand, the stalks are used or sold for fuel, the leaves are available as fodder, and being a hardy crop requiring little rainfall to sustain it, it is grown largely in a year of drought. The other miscellaneous food-grains require no detailed description. The chief are peas, *masurī* (*Ervum Lens*), *kodo* (*Paspalum scrobiculatum*), *chīna* (*Panicum miliaceum*), *sāwān* (*Panicum frumentaceum*), *urid* (*Phaseolus Roxburghii*), *mūng* (*Phaseolus Mungo*), *janerā* (*Holcus sorghum*) and oats. All of these, except oats, are generally sown with other crops. Their value as food-grains is not high, and they form, as a rule, the poor man's food. *Kodo*, *chīna* and *sāwān* are boiled and eaten like rice. *Masurī*, *urid* and *mūng* are consumed as pulses. *Janerā* and oats are largely fodder crops. The whole of these minor food-grains cover less than one-fifth of the net cropped area of the district.

Maruā. The most important single crop in the district after rice is *maruā* (*Eleusine Coracana*), which covers 13 per cent. of the district area, a proportion which is fairly equally maintained in each subdivision. In Samāstipur this crop is rather less extensively grown than in the northern part of the district; and it is most prevalent in the Khajaulī and Madhubanī thānas. It is the staple crop of a considerable part of the district, and is cultivated with considerable care. It is a cheap grain, mainly eaten by the poorer classes. The grain is pounded into flour and consumed in the form of cakes, while the stalks are useful as fodder. *Maruā* is the only crop which is transplanted like rice; it is reaped in September, and the land on which it is grown almost invariably produces a second crop at the *rabi* harvest.

Barley. The next most important crop in the district is barley, which covers about 9 per cent. of the cropped area. It is grown on fairly high lands, which have generally already produced a *bhadoi* crop. It is largely consumed in the form of cakes by the poorer middle classes, who cannot afford rice or maize.

Maize. After barley comes maize or Indian corn (*makai*), covering 5 per cent. of the district cropped area, but it owes its position to its predominance in the Samāstipur subdivision, where it occupies 14 per cent. of the net cropped area and is the crop next in importance to rice. It is often sown with *rabi* crops, and even where this is not the case, a *rabi* crop is usually sown in the land after it has been cut. The crop is often very luxuriant, especially when grown on the high lands round the village homestead. It is used as food in various forms. By the well-to-do, the cobs are often eaten roast or fried, but usually they are threshed and ground into flour, and then eaten either as *sattu* or in the form of cakes.

Wheat. Wheat covers only 3 per cent. of the net cropped area, and is least important in Madhubanī and most prevalent in Samāstipur. It requires a rich and fairly dry soil, as it is very subject to blight. It is often grown as a second crop in the best *makai* lands, but in many cases wheat lands are left fallow during the rains. It is a valuable crop, and but for its want of hardiness would probably be more widely grown than it is. As a food, it is eaten by the well-to-do classes.

Gram. Gram (*Cicer arietinum*), the last of the important cereals, covers only a little over 2 per cent. of the cropped area It is most widely grown in the Darbhangā thāna, where, as elsewhere, it is often sown as a catch-crop in winter rice lands. It is usually eaten in the form of flour or *sattu*.

Non-food crops are grown on 15·79 per cent. of the net cropped area. They predominate in the Samāstipur subdivision, where they occupy 22 per cent.; while in each of the other two subdivisions they occupy only 14 per cent of the total cropped area. In Samāstipur, the major portion of the large non-food crop area is devoted to indigo and tobacco, two very valuable crops, which are unimportant elsewhere. On the other hand, the area under sugarcane is mainly found in the north of the district, particularly in the Madhubanī subdivision, where it has recently been substituted largely for indigo. But, as a whole, the more valuable non-food crops are grown mainly in the Samāstipur subdivision. **NON-FOOD CROPS.**

Oil-seeds are the most extensively grown non-food crop in the district. They consist mainly of linseed, but comprise also 25,000 acres of mustard, and small areas of *til* or gingelly (*Sesamum indicum*) and castor-oil. They constitute, however, the least valuable non-food crop, being most extensively grown in the head-quarters and Madhubanī subdivisions as a catch-crop, after the harvesting of the winter rice. They cover nearly half of the non-food crop area, but their value is often inconsiderable. **Oil-seeds.**

A large area (29,800 acres) is under tobacco, which is probably the most paying crop in the district. Five-sevenths of the total area is found in the Samāstipur and Dalsingh Sarai thānas, and most of the rest is in Wārisnagar and Benīpatī. The latter, though mainly a rice-growing tract like the rest of Madhubanī, contains a considerable stretch of high land which is used for the more valuable *rabi* crops. The fame of the tobacco of *pargana* Saraisa, which covers a large part of the Samāstipur and Hājīpur subdivisions, has spread far beyond the limits of North Bihār, and every year purchasers come from distant districts and buy large quantities of it. **Tobacco.**

The cultivation of poppy has never been widely practised in south of Darbhangā, its place being taken by its rival, tobacco; and the greater part of the opium grown in the district comes from the Benīpatī thāna. The area covered by it is not large (2,400 acres), but the price obtained for the crude opium renders it a very valuable crop. The production of opium is a Government monopoly, and no person is allowed to grow poppy except on account of Government. Annual engagements are entered into by the cultivators, who, in consideration of the payment of an advance, agree to cultivate a certain quantity of land with poppy and to deliver the whole of the opium produced to the Government at a rate fixed according to its consistence, but subject to deductions for inferiority of quality. The best soil for poppy is loam, so situated **Opium.**

that it can be highly manured and easily irrigated, and for this reason homestead land is generally selected. The cultivation requires much attention throughout the growth of the plant. From the commencement of the rains in June until October the ground is prepared by repeated ploughings, weedings and manuring, and the seed is sown in November. Several waterings and weedings are ordinarily necessary before the plant reaches maturity in February. After the plant has flowered, the first process is to remove the petals, which are preserved, to be used afterwards as coverings for the opium cakes. The opium is then collected during the months of February to March, by lancing the capsules in the afternoon with an iron instrument and scraping off the exudation the next morning. In the beginning of April the cultivators bring in their opium to the weighment centres of the different Sub-Agencies, where it is examined and weighed, and the balance due according to the Opium Officer's valuation is paid to them. Final adjustments are made in August after the value of the drug has been ascertained by assay at the Patna Factory, where the final process of preparing the drug in balls or cakes is conducted.

There is a tendency for the cultivation of poppy to decrease, as year by year it is becoming less profitable to the ryots. The plant is delicate; a thoroughly favourable year comes only at uncertain intervals; and the cultivators have had to contend with a number of bad seasons. There is accordingly a marked tendency to withdraw from an industry so precarious, and to substitute the more robust cereals or such paying crops as sugarcane, potatoes, chillies and vegetables. This movement has been quickened by the fact that the value of cereals has increased of recent years, while the price paid for the crude drug remains stationary.

Indigo.

Indigo is most widely grown in Samāstipur, but was until recently very generally cultivated in all parts of the district. Its cultivation has been practically abandoned in the three northern thānas of Madhubanī, and within the last few years the area under the plant has shrunk enormously. It is the most valuable of all the non-food crops, but as it plays such an important part in the rural economy of the district, an account of its cultivation must be deferred to Chapter IX.

Sugarcane.

The most important crop after indigo is sugarcane. It has long been grown in the Madhubanī subdivision and in Baherā thana, but its cultivation has been extended of recent years owing to the decline of indigo. In earlier days the cultivation of sugarcane was very considerable, and many of the indigo concerns were originally started as sugar factories, the manufacture of sugar being given up when indigo proved to be more profitable. The reverse process is now taking place, as indigo is in many places being replaced by sugarcane. In the recent settlement operation it was found that the area under sugarcane was 16,300 acres or 0·96 per cent. of the net cropped area, but it has now increased to 17,000 acres. Sugarcane is one of the profitable crops grown in the district in spite of the labour and expense its cultivation requires. It is a crop which not only exhausts the soil, but occupies the ground for a long period, extending over a year. It is planted during February or March, in cuttings of about a foot in length placed in rows about 2 feet apart. When the plant begins to sprout, it is well watered and the surrounding earth is loosened. Each plant grows into a cluster of canes, which are generally ready for cutting in January or February. The crop requires great care, and must have 7 or 8 waterings, even if the other crops have to do without water in consequence.

As the price of sugar fluctuates considerably, its cultivation is always somewhat of a speculation. One attraction, however, to the ryot is that he can generally get an advance from the person for whom he grows the crop, and this factor almost invariably causes an extension of sugarcane cultivation in a year of drought, when other crops have failed and the cultivators are pressed for ready money.

Other crops.

The only other non-food crops which requires special mention are fibres and the thatching grass called *kharaul*. The latter grows wild, if left undisturbed, and, like sugarcane, shelters herds of wild pig; when green, it affords fodder to the cattle, and when dry, it is used for thatching. If only cattle are prevented from trespassing, it is as profitable as many of the food crops. The area under fibres is insignificant, and is mostly accounted for by cotton cultivation in the Samāstipur subdivision. Experiments with tree cotton have recently been started at Meniārpur, and with rhea at Pandaul and at Dalsingh Sarai, 500 acres having been planted out at the latter place.

Fruits and vegetables.

One of the most remarkable features of the district is its numerous and extensive mango groves. They are to be found in almost every village, and altogether cover 85,000 acres or 4 per cent. of the area under cultivation. The trees generally are well grown and of uniform size, being planted with great regard for symmetry, in straight equidistant lines. The *lichī* (*Nephelium litchi*) gardens are justly famous for the delicious fruit they bear; peaches and limes flourish, and among other fruits grown with success may be mentioned the *papaya*, plantain, guava, custard-apple, jack-fruit and *bel*.

Among vegetables, potatoes, yams (*aluā*) and *suthni* are very common, especially in the Samāstipur subdivision, where they occupy a large area. A great number of other vegetables are raised in garden plots for household use, and also on a larger scale in the neighbourhood of towns, such as the egg plant or *baigun* (*Solanum melongena*), ground-nut (*Trichosanthes dioica*), pumpkins, gourds, radishes, melons, onions and carrots. The favourite condiment is the chilli, which is cultivated extensively.

EXTENSION OF CULTIVATION. A century ago a great part of Darbhangā was uncultivated partly owing to the desolation caused by the terrible famine of 1770, and partly because of the oppression of the farmers of revenue and the depredations of freebooting zamīndārs. So terribly did the former affect the people that in 1783 the Collector of Tirhut submitted a proposal that cultivators should be attracted from the dominions of the Vizier of Oudh to reclaim the unpeopled wastes of his district; and in 1781 the Judge reported that, owing to the tyranny of the local revenue officer and his subordinates, there was but very little cultivation for 20 miles from Darbhangā and that grass jungles appeared over extensive plains which before were rich in culture. In 1796 *pargana* Pachhi was described as the abode of dreadful beasts of prey, while the adjoining *pargana* of Alāpur, now one of the richest parts of the district, was the haunt of wild elephants, whose depredations prevented all improvement. *Pargana* Bharwāra, which comprises a considerable portion of the head-quarters subdivision, where 78 per cent. of the land is now under cultivation, contained large stretches of waste land; and in 1802 it was reported that for miles nothing could be seen but uncultivated plains with here and there a few *bighas* under the plough. When law and order were finally established, agriculture was extended to a remarkable degree, and with what rapidity this happened can be gathered from a report submitted by the Collector in 1824. "In Tirhut proper," he wrote, "the waste land at the time of the Settlement, it is believed, considerably exceeded that under cultivation, and in some extensive parganas adjoining the Nepal Terai and those between the Teljuga and the Daosi rivers in the north-eastern part of the district the cultivated land was to the waste, perhaps, as one to fifty. All these parganas are now considerably advanced in cultivation. Since the decennial settlement, cultivation has been improved and extended, and the population has increased in Tirhut in a manner that excites the wonder of those who do not reflect that such was the inevitable result of that measure."

In the early part of the 19th century then a very large proportion of the district, amounting probably to half the total area, and in the north certainly to more than half, was uncultivated. By 1840 the cultivated area had increased to three-fifths of the total, but in the north it was still little more than half. During the next ten years, the cultivated area appears to have increased rapidly, until in 1850 it amounted probably to nearly three-fourths of the total. In 1875 it was estimated to be 79 per cent. of the total; and the recent survey and settlement operations have shewn that it is now just under 80 per cent. of the total area of the district. We should probably therefore be justified in concluding that cultivation has nearly doubled itself within the last hundred years, but that the greater part of the increase took place in the first half of last century.

The largest uncultivated area is in the head-quarters subdivision, where there is a considerable amount of swamp and marsh, which is under water for the greater part of the year. It is nearly as great in the Madhubanī subdivision, where there is much culturable jungle along the banks of streams and on the Nepāl frontier; and it amounts to 23 per cent. of the total area in the Madhubanī thāna, where it is due to the large number of mango groves which stud the country; this tract is the stronghold of Brahmanism in Darbhangā, and many of the groves have been planted from religious motives. It is least (17 per cent.) in the Samāstipur subdivision, where there is less room for the extension of cultivation than in any other part of the district. It may appear somewhat remarkable that Samāstipur under these circumstances shows a higher proportion of current fallow than Madhubanī, but the reason is that the agricultural conditions of the two subdivisions are radically different. Samāstipur, as a whole, is marked by the careful cultivation of its rich uplands. Such lands require occasional rest, and the cultivators are intelligent enough to understand the advantage of allowing it to them. But in Madhubanī the lands are mainly low and produce but one crop in the year, and so the necessity for high cultivation does not arise. In a year of good rainfall no one would think of leaving rice lands fallow, and, indeed, rice lands positively deteriorate when left uncultivated, as they become baked and hardened, the *ails* or partitions between the fields become broken, and the drains by which the land was irrigated get filled up or obliterated.

The uncultivated area is one-fifth of the total area of the district; but about half of this is composed of roads, rivers, tanks, house-sites, etc., and is therefore not available for cultivation. Of the area still available for cultivation about half is waste land, pure and simple; and much of it, being impregnated with the salt efflorescence known as *reh*, is unfit for cultivation in its present state. But in a district such as Darbhangā, where the cattle mainly depend upon grazing, it would probably be impossible to reduce the area of waste land to any appreciable extent; for if the area still culturable were to be cultivated, there would be nothing left for the cattle to graze. There is thus but little room for the further extension of cultivation; the district is already densely populated; and there appear to be good grounds for the belief that it will at no distant date reach a point when it will no longer be able to support an increase in its population from the produce of the soil, without either a reduction in the standard of comfort or an increase in productive capacity.

IMPROVED METHODS OF CULTIVATION.

So far there appears to be no indication of any general improvement in the methods of cultivation. The implements are the same as have been used for generations past, except in the lands worked by European capital, and no new or improved implements have been adopted, with the exception of the iron roller sugar-pressing machines invented in 1874 by the proprietors of the Bihiā estate in Shāhābād, and hence known as the Bihiā miles. No greater progress is noticeable in the use of manures, except where indigo is grown and European enterprise has developed the land. Manuring is, indeed, little resorted to, when it involves any extra expenditure, except for special crops like potatoes, sugarcane, tobacco, and poppy, and then only to a limited extent.

CATTLE.

Tirhut has always had a high reputation for its cattle, and the East India Company used to obtain draught bullocks from it for the Ordnance Department. The best strain in the district is the Bachaur, so called from the northern *pargana* in which the breed is common. Further south, floods militate against success in breeding, and though there is never an absolute lack of food, the want of good pasturage is a serious drawback. There are large grazing grounds in the north, but in the south, where nearly all the land is under cultivation, the cattle have to be content with such scanty herbage as the road-sides, tank-banks and boundary ridges afford, and are partly stall-fed on chopped straw. The grazing ground necessary for the maintenance of a sufficient agricultural stock has, indeed, probably been reduced to the lowest possible limits.

The Madhubanī subdivision is well stocked with cattle, and the Khajaulī and Phulparās thānas contain a large number of milch kine, draught cattle and agricultural stock of all kinds. These two thānas contain the Jabdi and Bachaur *parganas*, which are noted for their breed of cattle. The head-quarters subdivision and the Wārisnagar thāna are the worst stocked of all. The whole of this area is liable to inundation, from which the cattle always suffer most severely; and in addition, a large part of Baherā and Ruserā are badly off for roads, and but few carts are kept there. Samāstipur and Dalsingh Sarai are wonderfully well-stocked, considering the small proportion of waste land which they contain, but the high cultivation practised in this area no doubt necessitates the maintenance of a considerable stock.

Buffaloes are reared in considerable numbers, and the district has long been noted for its breed, good males fetching a high price and being largely exported to Bengal. They are chiefly valuable for the milk they yield in large quantities; the soil being lighter than in the South Bihār districts, they are rarely used for ploughing in this part of the country. Sheep are mostly found in the south of the district, but their number is inconsiderable. Goats are bred in most villages, and are especially numerous in localities in which Muhammadans form a large proportion of the population. They are most common in the Wārisnagar and Samāstipur thānas, owing to the number of Muhammadan villages in the neighbourhood of Samāstipur town. Pigs of the ordinary omnivorous kind are kept by the low castes. The only horses in general use are the usual indigenous ponies; they are generally undersized and incapable of much heavy work; but they are very hardy, and those used for *ekkās* often have astonishing endurance and a great turn of speed.

Veterinary institutions.

The diseases most prevalent among the cattle are rinder pest and foot-and-mouth disease; in 1903-04 over 2,750 cases of the latter disease were treated by itinerant Veterinary Assistants in the villages in the interior. Veterinary aid is also afforded at the Veterinary Dispensary at Laheriā Sarai. The people are beginning to appreciate the usefulness of this institution; and in 1905-06, 60 horses and ponies and 40 cattle were treated as in-patients, while the number of out-patients was 300 and 600 respectively.

CHAPTER VI

NATURAL CALAMITIES.

FLOODS. AN account of the river system of Darbhangā has been given in Chapter I, from which it will be seen that the district is intersected by numerous streams and rivers, many of which rise in the Nepāl hills and are subject to violent floods. To the north the Kamlā runs due south from the hills, being joined on the way by numerous tributaries, and pursues a southerly course as far as the town of Darbhangā; it then turns off to the south-east, and eventually falls into the Ganges. Further south the district is traversed by the Bāghmati, which rising near Kātmāndu, pierces the Nepāl hills before it reaches British territory. It receives many tributaries in the Nepāl *Tarai*, and after forming the boundary between the Muzaffarpur and Champāran districts, deflects abruptly to the south-east through Muzaffarpur and Darbhangā, until it joins the Tiljūgā at Tilakeswar in the south-east corner of the district. The Little Gandak rises in the lower range of the Nepāl hills near Sumeswar, and receives on its left bank numerous tributaries before entering North Bihār. It pursues an easterly direction through Darbhangā almost parallel with the Bāghmati, and, after traversing the southern part of the district, falls into the Ganges just below the town of Monghyr.

In this portion of their course the beds of the rivers, raised by the silt they bring down, are elevated above the level of the surrounding country; and it will thus be understood that sudden heavy rain over the northern portion of Tirhut, the Nepāl *tarai* and the range of hills to the north of it, some 150 miles in length, may easily cause these narrow drainage channels to overflow their banks. When this happens, an inundation inevitably follows, and the lower down one goes, the more is the evil aggravated, not only by the converging of the streams towards the same point, but also by the network of private *bāndhs* or embankments that have been made year after year for the protection of the villages along their banks: for in the first place they make matters worse for the villages lower down stream, and in the end, when the floods have asserted themselves, and spread over the country, they hedge them in and delay their subsidence. Owing to this combination of circumstances, the district has always been subject to severe and widespread inundations, which cause a good deal of temporary suffering. But, as a rule, the distress they cause soon passes away; the dwellings which are destroyed are quickly replaced, as the cost of erecting such mud-walled huts is small; and the cultivators are compensated, in large measure, for the losses they sustain by the fertilizing silt left by the receding waters, which increases the productiveness of the soil and ensures rich crops. The following account of the greatest of the floods which have occurred in recent years will sufficiently show how short-lived are their injurious effects, except in years of extraordinary floods such as 1906.

In 1893 there were three distinct floods, one after the other, answering to the heavy falls of rain in July, again in August, and lastly in September. The first did no particular damage, but the two others came down upon a country already water-logged, causing an immense amount of injury to crops, houses, roads, and to the railway line between Samāstipur and Sītāmarhi. Most of the *bhadoi* over the flooded tracts was lost, the portion saved being greatly injured; the rice plants were swept away not only once but twice, and in some places three times in succession, where the people had been able during the intervals to procure new seedlings and to replant them. In Champāran and Muzaffarpur the floods from the Bāghmati and Little Gandak were confined, roughly speaking, to the country lying to the north of the Bengal and North-Western Railway line. But in Darbhangā matters were aggravated by the water from the Kamlā, at its highest known flood, heading the flow from the other districts. The combined floods swept the district from north-west to south-east at a level some 3 feet higher than had ever been known before. One-half of the district was for the time a slowly moving inland sea, the inhabitants with their cattle camping out along the high-level roads, the railway embankments, and here and there on a piece of *dih* land standing out of the water like an island. Fortunately, the waters rose gradually; no lives, so far as could be ascertained, were lost; and the people had time to save their stores of grain and to drive off most of their cattle. Still numbers of these perished both during the floods and subsequently from want of proper nutriment.

Floods of 1893.

Altogether 2,000 square miles were submerged, the parts most affected being the centre and south of the head-quarters subdivision, the western portions of the Madhubanī subdivision, and the northern and south-eastern parts of the Samāstipur subdivision. The southern half of the town of Darbhangā itself was for some days under water, and between 1,400 and 1,500 houses collapsed. Great damage was done to the railway line, roads and bridges, traffic on the Darbhangā-Sītāmarhi section of the railway being interrupted for nearly three weeks. The way in which the people recovered from their losses, instead of being overwhelmed by them, was very remarkable. At first, it seemed almost certain that the affected tracts would be the scene of famine relief on an extensive scale; but as time went on, and the floods subsided, the people returned to their villages, rebuilt their houses, and proceeded to till their lands for the *rabi* sowings. Provision was made for distributing advances under the Agriculturists Loans Act; but, in the end, only a comparatively small sum was needed for the purpose. In February 1894 test relief works were opened in the south-east of the district, but they were attended by only 700 or 800 persons, and they were soon closed, as even this small number soon dwindled down. The reason for this appears to be that there were excellent harvests in the unflooded parts of the country, where there was a plentiful demand for labour, and prices were low; and that in the flooded parts the people expected an abundant *rabi* harvest after the enrichment of the lands by the silt deposit, and the *mahājans* were consequently willing to loosen their purse-strings.

Flood of 1898.

In September 1898 the district was again visited by flood, and an area of 800 square miles was inundated. In the Madhubanī subdivision the floods affected only the western portion comprised in the Benīpatī thāna, where they were caused by the rising of the Little Bāghmati. In the head-quarters subdivision the tracts of country along the Kamlā, Bāghmati, Little Gandak and Karai suffered considerably, but in the interior little damage was done, except to the low lands. The civil station of Laheriā Sarai and the town of Darbhangā, lying between the Kamlā and Bāghmati, were flooded in parts by the rise of both rivers, and about 800 houses collapsed; but the land on which the Government buildings stand was not submerged, owing to the protection afforded by sluice gates erected some time previously. In the Samāstipur subdivision the parts most seriously affected were the eastern portion of the Wārisnagar thāna along the Bāghmati and the south-western part of the Dalsingh Sarai thāna, which was flooded by the overflow of the Tāl Barailā, a large lake in the Muzaffarpur district. Fortunately, the floods came after the middle of September, when a considerable portion of the *bhadoi* crop had already been harvested; and, in spite of the loss caused by the floods, an average crop was secured. The winter rice crop in some places was entirely ruined, and altogether about half of it was damaged. The mud-walled huts of the villagers collapsed in great numbers, and over 38,000 were destroyed. Very few cattle were lost, but 164 persons were drowned, the loss of life being greatest in the Wārisnagar thāna, where the Bāghmati came down in flood with great suddenness, accompanied by a violent cyclone. On the other hand, the floods greatly increased the fertility of the soil, and ensured a magnificent *rabi* crop. Labour was abundantly provided for on the construction of the railway line to Katihār, which was close to the tracts most severely affected; and not a single application was made for Government loans. Prices did not rise, and the Collector reported that, taking the district as a whole, the flood was rather beneficial than otherwise.

Flood of 1902.

Another great inundation was that of August 1902, which was caused by excessive rain in the Nepāl hills. To the north-west, in the Benīpatī thāna, the Jamunā and Bilautī poured down their flood waters through the Dhaus river, and further east, in the Madhubanī thāna, the floods came down through the Jaynagar Kamlā, the Mirzāpur Kamlā, the Balān and Tiljūgā. The *bhadoi* crop was partially destroyed, the roads were extensively breached, and parts of the town of Madhubanī were submerged. To the north-east of Madhubanī the floods came down through the Balān and Bihula, joining the Khurg and Panchi channels; and in this tract the rice crop was seriously damaged. The flood was greatest in the Madhubanī subdivision, but further south, the Little Bāghmati breached the embankment which protects Darbhangā, and flowed into the southern and western parts of the town. Some injury was done to the crops in the headquarters subdivision, and also in some villages on the banks of the Little Gandak in the Samāstipur subdivision; but in this part of the district, the floods caused comparatively little damage, as they had already spent their force in the northern tracts. Even there, however, there was but little loss of life, 17 persons only being drowned. In some of the frontier villages many houses were damaged or swept away altogether, and there was considerable loss of cattle; but elsewhere the news of the flood spread very fast, and the villagers saved their cattle by pegging them out on the banks of tanks and on the roads. Considerable injury was done to the roads; and a bumper *bhadoi* crop was converted into a crop estimated at 78 per cent. in the tracts not flooded to less than half a crop in the flooded tracts. As, however, the flood occurred early in the season, the winter rice did not suffer much, as it was possible to transplant much of it, and the plentiful moisture enabled the cultivators to do so over a larger area than usual. In spite of the losses they sustained, the cultivators in the north welcomed the floods. In the preceding year the rainfall had been deficient, and the winter rice crop had nearly failed in some parts. Here the floods were of comparatively short duration, but they sufficed to replenish the empty tanks and wells, and left a thick deposit of silt, which was invaluable for the *rabi* crop and ensured a full winter rice harvest, where seedlings were available. Consequently, though the people complained of the embankments and of the passage of the flood, no complaints were made about the occurrence of the flood itself.

Floods of 1906.

It has hitherto been a common belief that in Darbhangā famine is never caused by flood, but this belief has been rudely shaken by the course of events during the present year (1906), when the district was swept by the most disastrous floods of which there is any record, and the suffering of the people culminated in famine. In the middle of July the river Kamlā rose in flood, causing considerable damage to the maize and *marua* crops, and also destroying the paddy seedlings in the headquarters subdivision; but the flood subsided after doing a comparatively small amount of damage. It was followed, however, by an inundation of unprecedented height and duration in August. This flood began to rise on the 6th August coming from two directions, the Kamlā and the northern or Little Bāghmati on the north-west, and the true Bāghmati and Little Gandak on the west. It rose steadily till the 24th August or for a period of 16 days, and during this time the greater part of the district was submerged. The whole town of Darbhangā and Laheriā Sarai was under water, the only places that escaped being the *kachahri* premises in Laheriā Sarai and the Bara Bazar in Darbhangā. Here the flood rose so suddenly and the rush of water was so great that thousands were rendered homeless within a few hours, and shelter had to be given to them in the *kachahri* compound. In the interior the distress was far greater. The mud-walled huts soon crumbled away, and for some days the homeless villagers had to camp out on any high land that stood above the waste of water, before they could move on to take shelter in the towns or the few villages that had escaped. The roads were broken in all directions, many parts were inaccessible for want of boats, the railway lines were breached in several places, and the bridges near Hayā Ghāt and Kishanpur being washed away, traffic was stopped for a month. The water in the town of Darbhangā began to recede after about a week, and had disappeared in about 2 weeks' time; but in the interior it took about 2 months for it to clear off, and in the meantime incalculable damage was done to the crops. The *bhadoi* crop was almost entirely destroyed in the head-quarters subdivision, and the damage done to it in the Madhubanī subdivision and the Wārisnagar thāna of the Samāstipur subdivision was scarcely less serious. The *aghani* crop, on the success of which the people are dependent for their year's supply of food, was practically swept away in three-quarters of the Darbhangā subdivision, and was nearly all destroyed in the Wārisnagar and Dalsingh Sarai thānas. In the north of the district, where the floods subsided early and seedlings were available, the cultivators were able to retransplant; but in the greater part of the head-quarters and Samāstipur subdivisions the lateness of the flood made this impossible.

The loss of the crops followed on a year of bad harvests, for in 1905-06 the *bhadoi* crop yielded only 40, and the *aghani* crop 67 per cent. of a normal crop; and the distress which ensued was aggravated by the high price of food grains. During the month of September the trade with adjoining districts was almost at a standstill owing to breaches on the railway line, and internal trade was paralyzed by breaches on the road. Prices had been gradually rising during the year, and were already very high; but the damage to the crops and the interruption in the communications into and through the district sent them up with a bound, until they reached even a higher level than in the famine of 1896-97. The price of rice rose to 5 seers a rupee, and could not be obtained at all in many places, while the quantity of maize available was unusually small, only a few maunds being for sale even in the town of Darbhangā. The result of this combination of circumstances was to plunge a considerable proportion of the population into destitution, and there can be little doubt that had it not been for the promptness of the local officers and planters in starting kitchens for the distribution of food among the homeless and infirm, many deaths from starvation must have been occurred. Famine has been declared in the Rusera and Baherā thānas, and relief operations have been started. The number receiving gratuitous relief was 45,500 in the beginning of October, 19,000 at the end of November, and 15,800 at the end of December; while the number of those attending test relief works has risen from 1,900 to 14,500 and 32,400 in the same period. Government has advanced 4 lakhs for distribution as agricultural loans, and this large sum has already been expended; while the

Mahārājā of Darbhangā, who has undertaken the task of granting relief on his own estates, has set aside 5 lakhs to provide for loans and gratuitous relief. Nearly two-thirds of the entire district with a population of 1,883,000 is affected, the area being 2,079 square miles, of which 1,069 square miles are in the head-quarters, 590 in the Madhubanī and 420 in the Samāstipur subdivision; and it is estimated that famine will continue until April, though it is considered unlikely that more than 1 per cent. of the population will require relief on relief works.

FAMINES.

Early famines.

In the early years of British administration, hardly a year passed without the record of some natural calamity; in one year it was drought, in the next inundation, and in either case the people were hard put to it to withstand distress. The earliest famine of which we possess any detailed record is that of 1769-70, when one-third of the population of Bengal is said to have perished, and Darbhangā suffered like the other districts of Bihār. A serious drought followed in 1783, and advances had to be made for the relief of the cultivators, though the scarcity did not culminate in actual famine. The years 1787 and 1788 were no more propitious, as the country was inundated, the cattle died in large numbers, and the crops failed. Drought again caused scarcity in 1791, when all persons, except grain dealers, were arbitrarily prohibited from keeping by them more than one year's supply of grain. There was another drought in 1804; in 1806 the *bhadoi* crop was entirely destroyed by floods, and the *aghani* was threatened; and in 1809 there was considerable suffering owing to the failure of all the principal crops.

Famine of 1866.

Though great distress was caused by these failures of the crops, they did not culminate in actual famine; and the first great famine of the 19th century was that of 1866. In the north of Tirhut, the rice crops had failed partially in 1863, and even more so in 1864. In October 1865, with the prospect of an even more complete failure on the Nepāl frontier, and a deficiency of the rice crops in Madhubanī and the north of the present head-quarters subdivision, the prices of rice and Indian corn rose to three times the ordinary rate, and cases of suicide from starvation and of deaths from want of food began to be reported. The harvesting of the rice crop, however, temporarily supplied both food and wages, and prices at the end of 1865, and in the beginning of 1866, were somewhat easier. In February, prices again began to rise; the usual grain supply from Nepāl ceased; and as there was no demand for labour, the poorer classes suffered extremely, and began to desert their homesteads and migrate southwards in large numbers, while fires and grain-robberies for the purpose of obtaining food became very common. The period from April to June never affords much employment to labourers, and this class, having nothing to fall back on, was reduced to the greatest destitution. No organized system of relief was commenced till June, but eventually relief operations were undertaken at 9 centres, 4 in the Darbhangā and 5 in the Madhubanī subdivision. The suffering was greatest during July, August and September, the price of rice ranging from 7 to 5¼ seers per rupee. In the middle of August cholera broke out, and the people reduced by long privation, fell easy victims to the disease. In September the extreme pressure began to diminish, owing to the good *bhadoi* or autumn crops; and fortunately it never increased again, as the winter rice, in most parts, was a good crop. Still, in some places, famine, disease, and desertion had so debilitated or diminished the population, that half the land remained uncultivated, and relief operations had to be carried on till February 1867.

The next great famine occurred only 8 years afterwards. The year 1873 was the last of a series of three years which were marked by abnormal rainfall and generally unusual weather. The year 1871 was unusually wet; the following year was equally dry; while in 1873 the rainfall was deficient beyond precedent; the rainfall registered at Darbhangā in these three years was 79, 42 and 24 inches respectively. The winter rice crop of 1871 had been seriously injured by inundations, and the consequence was that in 1872 the prices of food-grains were considerably above normal rates, and it was not until the excellent rice harvest of the winter of 1872 that the market recovered its usual tone. The *bhadoi* harvest of 1872, the ensuing rice harvest, and the *rabi* harvest of 1873 were full average crops; and the result was that, although the previous dear year might have pressed on the resources of the people, there was as much grain in the country in the autumn of 1873 as there usually is at that time of the year. **Famine of 1874.**

The rains of 1873 commenced late, were insufficient to bring even the *bhadoi* crops to maturity, or to permit of the usual rice area being sown, and ceased in September with a deficiency under the normal fall varying from 11 inches in Tājpur (Samāstipur) to 23 inches in Madhubanī and 28 inches in Darbhangā. The inevitable consequence was the failure of the rice crops in every subdivision, culminating in the almost total destruction of the winter rice crop in Darbhangā, where the rainfall from May to November was only 21 inches. The failure of the *bhadoi* crops varied in different portions of the district, the outturn in Madhubanī and Tājpur being returned as half and in Darbhangā as five-eighths of an average crop. The failure of the winter rice crops was, however, even more complete than this; for in Tājpur only one-fourth, in Madhubanī three-sixteenths, and in Darbhangā one-eighth of an average crop was saved. As regards the *rabi*, the crop was five-eighths or perhaps even three-fourths of an average crop in the Tājpur subdivision, but in the two northern subdivisions the great drought of 1873 made it impossible to sow the lands.

The deficiency in the food supply was supplemented partly by private trade, partly by the grain imported by Government; but the combined efforts of both agencies failed to raise the stocks in the district to a level sufficiently high to obviate the prevalence throughout the year of general tightness in the markets, of prices altogether abnormal, and of continuous pressure on all classes. Private import trade reached its highest degree of development in Tājpur, where relief was consequently least required. On the other hand, trade exhibited but a flickering vitality in Darbhangā, and this only in the south of the subdivision about the mart of Rusera; and in Madhubanī, it never showed any signs of vitality from first to last. In Madhubanī and Darbhangā, therefore, Government was compelled to put forth its full strength in its most organized form, and there were not wanting anxious days in March and April, when even sanguine men doubted whether the task was not too great for the resources of Government and the devotion of its officers.

The distribution of charitable relief was commenced in the latter month and closed early in October of the same year, the average daily number gratuitously relieved being 124,000, of whom 81,000 were inhabitants of the head-quarters subdivision, 38,000 of the Madhubanī subdivision and 5,000 of Tājpur. The highest daily number for three subdivisions was 134,000 in July in the head-quarters subdivision, 71,600 in August in Madhubanī, and 7,500 in July in Tājpur. Labourers were relieved by wages in grain and cash from January to September 1874, the highest numbers reached being 220,000 in April in the head-quarters subdivision, 369,000 in May in Madhubanī, and 13,300 in June in Tājpur. In April the total number in receipt of this form of relief was 531,500, and in May 569,400, but after those months the number rapidly fell, till it was under 22,000 in September. The relief given by Government in various shapes was on a vast scale, and was estimated to be sufficient for the support for one month of 1,954,732 persons in the head-quarters subdivision, for 2,925,146 in Madhubanī and for 138,638 in Tājpur.

Scarcity of 1876. Two years afterwards the district had to contend against scarcity owing to the failure of the winter rice crop. This caused some suffering in the north-eastern portion of the district, and the administration of relief had to be commenced in February 1876. Employment was offered, to a small extent, on relief works such as tanks and roads; but the relief was chiefly in the shape of village charity to the weak and sickly, and the expenditure only amounted to Rs. 30,000.

Scarcity of 1889.

Two periods of scarcity then intervened before the great famine of 1897. In 1889 there was some distress in a strip of country along the Nepāl frontier which includes the Khajaulī thāna. In this area the rainfall was much lighter than elsewhere and ceased prematurely, with the result that the winter rice crop, which is the mainstay of the people, was entirely lost, while the *bhadoi* also failed in places, and the *rabi* was sown under unfavourable circumstances. Relief operations were commenced in January 1889, the people being employed on the excavation of tanks and also on the construction of the railway line from Darbhangā to Sītāmarhi. The number on relief works gradually rose until the first fortnight of June, when it aggregated 14,656; but after that, abundant rain having fallen, many were induced to return to their ordinary occupations. It rose again to 19,570 at the end of August, but then declined steadily, and the relief works were finally closed in the first week of October. The total expenditure on these works was 2¼ lakhs, of which half was borne by the District Board and half by Government.

Scarcity of 1892.

There was again scarcity in 1892, the great deficiency of the rains of 1891 causing a total failure of the winter rice crop on the high lands and serious injury to the *bhadoi* crops; the outturn of the latter was estimated at not more than 37½ per cent., and that of winter rice was about the same. The affected tracts were the Ruserā and Baherā thānas, and a small portion of the Darbhangā thāna, to the east of Darbhangā town, in the head-quarters subdivision; and the Benīpatī and Khajaulī thānas, the northern half of the Madhubanī thāna, and parts of the Phulparās thāna in the Madhubanī subdivision. Relief operations were begun in March, and were continued for 19 weeks, the average daily attendance being 19,200. Altogether 67 tanks were dug at a cost of Rs. 94,000, and 994 miles of road were taken in hand, of which 582 miles were completed at a cost of one lakh. The highest daily number of persons relieved (47,450) was reached in the middle of June, but the daily wage was then reduced, and the rain which fell at this time enabled the people to return to their ordinary occupations. This led to a speedy and steady decrease in the number of labourers, which continued until the works were closed at the end of July.

Famine of 1897.

The last great famine from which Darbhangā has suffered is that of 1897. The course of events in 1896 strikingly illustrated the principle that the seasonable distribution of the rainfall is of even more importance, within certain limits, than its actual quantity. There was virtually no rain in the cold weather of 1895-96, and as the September rain was not very heavy, while the rainfall in October was practically nil, the water level had sunk very low by May 1896, when there was fairly heavy rain: in May, however, as throughout the season, the rain fell in a few heavy showers which quickly dried up. There was again good rain in the end of June and beginning of July, but then commenced the first long break of about 6 weeks: hardly any rain fell during this period, while there were many cloudless days of burning sunshine. Thus the rain which fell in August found the ground very dry, and the water level little, if at all, higher than in June. Towards the close of August there were a few days of fairly heavy rain, and then another long break, with fine hot days and occasional west winds, till after the middle of September. This rain again fell on a dry and parched up soil, and in most cases it almost disappeared in a few days.

The monsoon rainfall was below the average, but it was not so much the small quantity of the rainfall as its bad distribution and the hot sunny intervals that had so disastrous an effect on the crop. Up to the end of July the deficiency was only 1¼ inch, but then the district suffered from two long periods of drought in August and September, and much damage was thus caused to the rice. In August the rainfall was nearly up to the average, but in September it was 3 inches short, and no rain fell after the middle of that month. The net result was a deficiency of 17 per cent. up to the end of October. The actual deficiency of rainfall was not so great as in some other districts, but it was so peculiarly ill-distributed during the critical months of August, September and October, that the rice on which the district so much depends was a very great failure. The short rain in September and October had a bad effect on the *rabi* crops, but fortunately good showers fell in the cold weather and the *rabi* which did come up was very good in the south, fair in Darbhangā and moderately good in Madhubanī. The *bhadoi* and winter rice, however, were very seriously affected by the short rainfall, not only as regards the outturn, but also as regards the extent of land under cultivation. The result was that though the *rabi*, owing to the winter rains, was as high as 15 annas, the *bhadoi* turned out a 10-anna crop, and the *aghani* was only 5 annas. But as the rice in an ordinary year supplies over more than half of the harvests of the district, it will be obvious that the loss of more than two-thirds of it meant great distress for the people, especially as the rice tracts lie close together, chiefly in one-half of the district, *i.e.*, the northern portion.

Commencing from the south where the Samāstipur subdivision lies, we find along the Ganges and south of the railway line as far as Samāstipur, and thence along the Burh Gandak river, a tract very similar to the southern part of the adjoining subdivision of Hājīpur, possessing a soil of great natural fertility, cultivated by expert and industrious husbandmen, and receiving in this year of drought nearly its normal rainfall: this tract, comprising the Dalsingh Sarai and Samāstipur thānas, was always treated as safe, and required no relief except in a few places. North of this fortunate tract a zone of slightly distressed country stretched across the district, being bounded on the north by the Bāghmati river. North of this stream the whole district was severely affected from the first, except isolated tracts which were exceptionally favoured; of these the most important were a tract to the west of the Tiljūgā river in the south of thāna Phulparās, a tract further north to the east of the same thāna, a strip along the Kamlā river from the northern boundary of the district down to Singiā, and a similar strip along the Karai river from Baherā down to Hirnī. In all these localities, except the last, which is a *rabi* country, the prosperity of the tract was due to irrigation from the rivers; other small isolated areas owed their partial immunity to better local rainfall or to their containing high land and *rabi* crops.

No fears of famine were aroused until the break of the rains became very prolonged in August, when the *bhadoi* began to wither, the seed-beds of rice were drying up, and large tracts of rice land remained unplanted. By the end of September it became clear that nothing but good rain in the *Hathiyā* asterism at the close of that month and the beginning of October could save the rice. These hopes proved delusive; the *Hathiyā* passed away without a drop of rain; and the rice was doomed. A test relief work was accordingly opened at Phulparās at the end of October, but this did not attract labourers and was very soon closed. About the middle of November there were signs that distress was beginning to appear among the poorer classes in the Madhubanī subdivision; and test works were opened at Kamtaul and Jaynagar, and the old test works at Phulparās were reopened. The number of labourers was however very small during this month, as the harvesting of the winter rice gave employment on much more remunerative terms than Government offered; and large number of labourers went into the Nepāl *Tarai* and the north of Bhāgalpur, where the rice crop was very good. By the end of November the signs of distress among the poorest classes had decidedly increased, and by the middle of December 5 works had been opened and the average attendance had risen to 713. From this period there was a very rapid increase, the number of works rising to 83 by the end of January 1897 and reaching the maximum of 199 by the 19th June. They then steadily decreased and had all been closed by the 25th September.

The number of labourers increased to 113,880 by the end of January and was 124,514 on the 24th April. It then fluctuated according as showers fell or not, reaching the maximum in the last week of May, when it rose to 145,116 or 6 per cent. of the population affected. In the 1st week of July there were 112,316 labourers, but after that the number fell week by week to 3,616 during the week ending the 11th September and to 250 during the week ending 25th September, when the works were finally closed. The distribution of gratuitous relief was commenced towards the close of November 1896, and at the end of December the average daily number relieved was 1,447. By this time the organization of circles began to take effect, and the numbers of those relieved by Government increased rapidly and without a break till the first week in March, when they had risen to 59,097. In the meantime, the Darbhangā Rāj had also commenced the distribution of gratuitous relief to its ryots, and the total number of those in receipt of this form of relief reached its maximum of 124,410 on the 21st June. It was still nearly 100,000 in the beginning of August, but after that it fell rapidly, and all gratuitous relief was stopped during the week ending the 25th September. The expenditure on this form of relief came to Rs. 4,10,819, some 25,000 persons on the average being relieved daily for 4 months, besides 86,690 persons who received gifts of money, grain or clothes once for all.

The total famine expenditure from public funds was Rs. 36,77,307, and the daily average number relieved was about 1,700,000, or a little over 6 per cent. of the population. There was very little private relief given, except by the Mahārājā of Darbhangā, who spent Rs. 2,45,500 on relief works and Rs. 1,22,800 on gratuitous relief: the aggregate number of those employed on these relief works from first to last, reckoned in terms of one day, was 2,834,848, and of those gratuitously relieved 2,943,702. The highest number of persons relieved on any one day during the famine was 253,910 or 10·49 per cent. of the population affected; and the total number of persons relieved, in terms of one day, was 40,911,000, a total larger than in any other Bengal district.

The death-rate was unusually low during the greater part of the distress. The only months in which the average of the preceding 5 years was exceeded were March, August and September, in the last two of which there was a great deal of fever: the excess was however largest in thānas outside the area of greatest distress, so that the rise in the death-rate can scarcely be attributable to famine. The recovery of the people was rapid, and the end of 1897-98 saw them almost restored to their normal condition.

Scarcity of 1906. In 1906 Darbhangā again suffered from scarcity, amounting in some tracts to famine, owing to the disastrous floods which swept the district. These floods and the acute distress they caused have already been described, and it will suffice to point out that the scarcity is entirely different from that which has visited Darbhangā hitherto in being due not to scanty but to excessive rainfall.

Liability to famine. Regarding the liability of the district to famine, the following remarks of the Settlement Officer, Mr. J. H. Kerr, I. C. S., may be quoted :—"It will be generally conceded that all famines in North Bihār, which have been serious enough to require Government relief on a large scale, have been due to the failure of the winter rice crop. There is, it is believed, no instance in which Government has had to undertake relief, save in limited local areas, owing to a failure of the crops of the *bhadoi* or *rabi* harvests, though, in the event of a failure of the winter rice crop, conditions are of course relieved or aggravated according to the nature of the preceding *bhadoi* and the subsequent *rabi* harvest. Thus it is only to be expected that Darbhangā, which has a larger gross area and a larger proportion of its cropped area under winter rice than any other district in North Bihār, should suffer most severely in the case of famine, and as a matter of fact, eight out of the ten thānas in the district, of more than four-fifths of its total area, have been officially declared liable to severe distress in the event of a failure of the winter rice crop. The statistics of the last great famine of 1897 further emphasize the importance of the winter rice crop in Darbhangā. The three districts of Sāran, Darbhangā, and Muzaffarpur are practically on an equality, so far as the pressure of population on the soil and the small area available for extension of cultivation are concerned. But the crop failure of 1896 affected them in different degrees of intensity. Though Sāran has the greatest density of population to the square mile and the highest rent rate, it suffered least severely from famine, owing to the variety of its crops and the practical independence of many parts of it of the winter rice harvest. In Darbhangā, on the other hand, where in two subdivisions winter rice covers over three-fifths of the net cropped area, more than a tenth of the population affected by its failure had to turn to Government for relief during the famine of 1897. There could be no more striking corroboration of the conclusion of the Famine Commission, that 'the devastating famines to which the provinces of India have from time to time been liable are in all cases to be traced directly to the occurrence of seasons of unusual drought, the failure of the customary rainfall leading to the failure of the food-crops on which the population depends.' And it might be added that in Bihār, famine, as distinct from local scarcity, is always due to the failure of the winter rice crop, and varies in its effects upon a given area with the importance of the crops other than the winter rice, on which the cultivators can fall back."

"As a matter of fact, the cultivators of Darbhangā can and do weather more than one season of crop failure without turning to Government for relief. It is a common proverb that it takes three bad years to make a famine. Even in the great famine of 1897, which followed two seasons of bad harvests and was accompanied by a higher range of prices than had been known since the previous famine of 1873-74, the proportion of pure cultivators who had to seek relief at the hands of Government must have been very small. For the total number relieved, when the distress was at its height was only 10 per cent. of the total population affected, or little more than half of those who are classed in the census returns as labourers, and who have either no land at all, or whose holdings are too small to support them from the profits of cultivation in an ordinary year, so that they have to eke out their income by working for others. It speaks well for the staying powers of the district, that a large minority even of the labouring class were able to dispense with Government relief during the last famine."

PROTECTIVE SCHEMES. Various projects have been put forward with the object of affording protection to the cultivators from the variability of the seasons. One is to establish irrigation works on the Dhaus, a minor river flowing from Nepāl into the north-western corner of the district, a tract which was severely affected in 1897. This scheme is however open to the objection that the Nepalese dam up the river, and that though the ryots might take the water in ordinary years, they would certainly object to paying for it. The supply of water is small, and it is liable to be cut off by the Nepalese, just when it might be wanted. Another and larger project is that of providing irrigation from the Kamlā by a main canal 12½ miles long, with three distributaries, which would command an area of 460 square miles. The cost of the project, which was first proposed in 1877, was estimated at Rs. 10,41,000 and the area likely to be irrigated at 52,500 acres. The Irrigation Commission (1901-03) recommended that a detailed estimate of this scheme should be prepared, and that if the cost per acre was not much greater than was anticipated, or not more than Rs 25 an acre, it should be sanctioned and put in hand as soon as funds could be made available. In doing so, they remarked:—" In view of the severe distress to which the densely populated districts of Northern Bihār are subjected whenever there is a failure of the autumn rains, and of the means of reliable protection that are available, we are unwilling to admit that the cost of protection will exceed its value until the matter has been put to a crucial test. Such a test the construction of the Kamlā project will afford at a moderate cost." Estimates are now being prepared; but the district seems to suffer nearly as much from flood as from drought, and it seems doubtful therefore if irrigation works with permanent head-works will be worth their cost. The project, the cost of which is roughly estimated at 14½ lakhs, will provide for the irrigation in a very dry year of about 30,000 acres of rice and 10,000 acres of *rabi*; but it is probable that in a moderately dry year the area of rice irrigable would be greater, while in an ordinary year there would be little or no demand for water. The country which would be commanded is liable to flood, and there are difficulties in finding a suitable site for permanent head-works; and, besides this, there is a risk of the river changing its course and leaving the head-works high and dry.

Another means of protecting the people from the effects of drought consists of the extension of the system of *pains* or private channels led off from the rivers, which has been instituted with such success by Mr. R. S. King in the Darbhangā Rāj. Regarding this method of protection, the Indian Irrigation Commission remark: " The excellent results show how much can be done, by active and energetic officers of long local experience, to utilize the available water-supply in seasons of drought by temporary and comparatively inexpensive expedients adapted to the exigencies of the moment. We think it almost certain that a great deal could be done at a comparatively small expense, and at a cost that would fall far below the net cost to the State of any ambitious system of permanent works, if in seasons of drought prompt measures could be taken for throwing earthen dams across the principal streams at the earliest possible moment, and for diverting the water through the network of channels already existing."

CHAPTER VII.

RENTS, WAGES AND PRICES.

Produce rents. Rents in Darbhangā are paid either in cash or in kind, but cash rents are of much greater importance, as they are paid for nearly four-fifths of the occupied area. Of the land held by under-ryots 53 per cent. pays rent in kind, but non-occupancy ryots pay such rents for only 7 per cent. of their land, while settled and occupancy ryots, who form the main body of cultivators in the district, pay them for no more than 8 per cent. of their holdings. As a rule, produce rents are most in favour where the proprietor is a petty resident landlord, who can attend personally to their realization, or where the land is liable to inundation or deterioration and the crop is consequently precarious. The largest area under produce rents is in the Dalsingh Sarai thāna, where petty resident proprietors predominate, the proportion being as high as 12 per cent. of the total ryotī area. After Dalsingh Sarai come the Darbhangā and Baherā thānas, with 10 per cent. paying produce rents; in these two tracts there are a large number of small resident landlords, and as much of the land is liable to inundation, the ryots in some villages prefer to pay a rent varying with the outturn rather than a fixed cash rent. In all the other thānas of the district, the area paying produce rents is below the district average, except in Madhubanī, where the peculiar conditions prevailing in the Madhubanī Bābus' estates raise the proportion to 10 per cent. These landlords maintain a very large establishment of Brāhman priests and other dependants, and have always found it convenient to realize a considerable share of their rents in kind.

The *batāi* system. Produce rents are of three kinds, *batāi*, *bhāoli* and *mankhap*. Under the *batāi* system, the actual crop is divided between the landlord and the ryot, either in the field or on the threshing-floor. The proportion is usually half and half, but in the north of the district it is not unusual for the landlord to take nine-sixteenths of the produce; the latter system is known as *nausātha*. The *batāi* method of division is usually very harassing to all concerned. The ryot is worried by the perpetual supervision of the landlord, and unless the latter can realize the grain in person, he is almost certain to be swindled by his agents. It may, indeed, be said that, with the exception of the *gumāshta* class, nobody benefits by the *batāi* system.

The *bhāoli* system. The *bhāoli* system, on the other hand, offers less opportunities for vexation or fraud. The value of the crop is appraised on the ground, shortly before the harvest, and a share of that value is paid by the ryot to the landlord, either in cash or in kind, as may be found most convenient. Its successful working depends on the manner in which the appraisement is carried out; but it is surprising with how little friction it is accomplished in the majority of cases. In some parts of the district, where the rice crop is exceedingly liable to inundation, it is probably as suitable a method of arriving at a fair rent as could be devised.

The *mankhap* system. So much cannot, however, be said for the *mankhap* system, under which the ryot has to pay certain quantity of the outturn, usually from eight to ten maunds a *bīgha*, irrespective of the amount of the produce. The rate is about double that of the ordinary cash rental, and though in a good year the ryot who pays *mankhap* may secure a larger share of the produce than the man who pays *batāi* or *bhāoli*, the *mankhap* system is intensely unpopular among the ryots. It deprives them of the benefit of high prices, which is the only compensation they receive for a season of short outturn, and cases have actually been known in which ryots, in a year of famine, had to buy grain at exorbitant rates to make over to their insatiable landlords.

CASH RENTS. The average rates of rent per acre paid by the several classes of ryots and under-ryots of Darbhangā are as follows:—ryots at fixed rents Re. 1-12; settled and occupancy ryots Rs. 3-12-6; non-occupancy ryots Rs. 4-7-10; and under-ryots Rs. 4-8-4. Ryots at fixed rates occupy a very advantageous position in Darbhangā, and the rents they pay are very low, especially in the north of the district. The highest rates paid by non-occupancy ryots are found in the Samāstipur subdivision, and the lowest in the Phulparās thāna, where their rents are lower than those of occupancy ryots. The rate payable by under-ryots is only 20 per cent. in excess of that payable by occupancy ryots, but it must be remembered that more than half the area held by the former pays a produce rent; and that, like the ryots at fixed rates and the non-occupancy ryots, they are only a small body, holding an inconsiderable area. In the case of settled and occupancy ryots, who hold over 98 per cent. of the ryoti area, the average rent-rate exceeds Rs. 4 per acre in only four thānas, Samāstipur, Dalsingh Sarai, Puserā and Khajaulī. In Samāstipur and Dalsingh Sarai, agricultural conditions have long been highly developed; there are many petty resident proprietors; and as the tract is highly productive and practically immune from the effects of drought, high rent rates are only to be expected. But neither Rusera nor Khajaulī are specially favoured in any respect, and the high rent rates prevalent in them can only be ascribed to the action of their chief landlords. The greater part of Khajaulī, which shows absolutely the highest rent rate in the district, belongs to the Rājnagar Circle of the Darbhangā Rāj, where the ryots are particularly subservient, and have allowed the rent rates to be forced up in a remarkable manner.

In the district, as a whole, the average incidence of rent to the occupied area is Rs. 3-12-4, varying from Rs. 3-10-9 in the head-quarters subdivision to Rs. 3-12-2 in Madhubanī and Rs. 3-15-4 in Samāstipur. These rent rates are deduced from the rents recorded as payable during the last settlement operations, and not from the rents claimed, and in some cases collected, by the landlords, which were often much higher. But even so, rents in Darbhangā have not yet become true competition rents, for they do not vary with the pressure on the soil. The highest rent rates are found in the Khajaulī and Ruserā thānas, where the population is only 766 and 763 to the square mile respectively, the reason being that in this tract big and powerful landlords have been able to make their tenants agree to periodical enhancements. On the other hand, many extremely congested villages in Samāstipur show very low rent rates, either because the ryots have been strong enough to resist enhancements, or because the landlords have not thought it worth while to enforce them. Illegal enhancements have, however, been made with such ease in all parts of the district that there can be no doubt that without a record of rights they would have become more and more common, rents would have tended more and more to become purely competition rents, and the provisions of the law restricting enhancement by contract between landlord and tenant would have become very largely a dead letter.

Enhancement of rents. The last Settlement Report shows that rent rates have already been doubled within a century; and from the road-cess returns it appears that in the 17 years between 1876 and 1893 the rental which the landlords claimed as payable had increased by over 25 per cent., though there could have been but little extension of cultivation during that period. The rate of increase is more than double that allowed by law in contracts between landlords and tenants; and the enhancement was obtained through the Civil Courts in less than one case in 10,000. Rent rates have, in fact, been subject to constant enhancements, both directly and through the imposition of illegal cesses and *ābwābs*. They were often made quite capriciously and without any regard to the capabilities of the soil, and were mostly effected by *sharanāmas*, or agreements executed by the leading ryots of the village, admitting that henceforth the rates were to be so much. It may be said that the *sharanāma* system is one of the most objectionable methods of enhancement. Legally, it only binds the people who are parties to the *sharanāma*, and then only if the resultant enhancement does not exceed the limits prescribed by law. But, practically, if the leading ryots of the village can be induced to agree to the *sharanāma*, the others soon fall into line; and as the agreement is only concerned with rates, it leaves the individual ryot in ignorance of his new rent until this is worked out for him by the *patwārī* or some other underling.

Many landlords do not even take the trouble to get *sharanāmas* executed when they want enhancements. They simply enter the new rent in the *jamābandi*, and exert pressure, both legal and illegal, to compel the ryots to pay it. One fruitful source of such enhancements has been the extreme subdivision of property by partition. The result has been the creation of a large number of petty landlords, who are unable to keep up their position, but who try to do so by squeezing as much as possible out of their tenants, and by arbitrarily enhancing their rents. "There can be no doubt," writes Mr. Kerr in the Darbhangā Settlement Report, "that the provisions of the Tenancy Act, limiting the enhancement of rent by contract, have hitherto been flagrantly disregarded or rather ignored. What the ryots objected to was not that the rents were such as they could not afford to pay, and still less that they were illegal (for they were absolutely ignorant of their legal rights), but that constant changes in the rent roll destroyed all security. In the district, as a whole, the canker which has mainly affected agricultural prosperity, is the constant tampering with rent rolls and consequent uncertainty as to rents. If our operations put a stop to this, they will have conferred no mean benefit on the district."

WAGES. Statistics of the wages paid for certain selected classes of labour and the rates current during the decade 1893—1902 will be found in the Statistical Appendix; and it will be sufficient to observe that higher wages than those quoted are obtainable in the railway workshops at Samāstipur, where labourers get

2½ annas, blacksmiths 7 annas, and carpenters and artizans 8 annas a day. It is interesting to compare these figures with those of 1794, when a common cooly received a wage of 1 anna 2 pies and a carpenter 3 annas per diem. Even 30 years ago a labourer received only 1½ anna a day, while the wages of smiths and carpenters were no higher than 2 to 4 annas, though they sometimes got 5 annas for the day's work. Apparently, however, the rise in the rate of wages in rural areas has not kept pace with the general rise in the price of food-grains, especially for the lower classes of labour. The village labourer earns about the same year after year, but in the towns, where there is a special demand for it, skilled labour commands a higher price than formerly. Among masons, carpenters and blacksmiths the wage shows an upward tendency; the silversmith charges a higher rate for his workmanship; the shoe-maker and the tailor have raised their tariff; and there is a similar tendency among domestic servants. The rise is small and gradual, but is observable all the same; and it appears to be due to a combination of circumstances, such as the advance in the standard of comfort among natives of the better class, the development of roads and railways, and the increasing communication with large centres of industry.

Outside urban areas the wages of labour maintain much the same level from year to year. Fortunately, however, wages in the villages are usually paid wholly or partly in kind; even the village artisan receives grain for the services he renders; and the field-labourer generally gets the whole of his wage in one or other of the inferior grains such as coarse unhusked rice. An able-bodied labourer gets 4 seers of the cheapest kind of grain available, such as *maruā*, gram or barley *sattu*. The money value of this works out to just under 2 annas, and the latter sum may be taken as the average. Women and boys get 3 seers only, equivalent to about one and a half annas. The wages of labour depend on the kind of work to be done. Heavy work, such as transplanting and digging, is always done by men, and lighter work, like weeding, is largely done by women. Reaping and threshing are also invariably paid for by a share in the produce, usually one-eighth; for some crops, however, which are easy to reap, like *maruā*, the proportion actually paid is a good deal less than this. When the cultivator does not watch his crop himself, he gives the produce of 5 *dhūrs* in the *bīgha* to some one to do it for him, and the same amount is paid to the blacksmith and carpenter for keeping his ploughs and *kodālis* in order. The *barāhil* and *gorait*, whose duty is to summon the ryots to the landlord's office for payment of rent or other purposes, are usually remunerated by being allowed to hold a small area of land rent-free. Similarly, the village tradesmen are nearly always paid in grain; and the washerman and barber get the produce of five or six *dhūrs* of land annually for each person for whom their services are required. This system is particularly suited to an agricultural district like Darbhangā, as it has the advantage of being unaffected by any rise in the price of food grains. Whatever the fluctuations in the price of these in the market, the labourer's wage remains the same, and he is thus to a certain extent protected against the distress caused by the high price of food.

A statement of the prices current in each subdivision during the years 1893–1902 is given in the Statistical Appendix. They show a great advance on those obtaining even a generation ago, for during the ten years 1861—1870, the average price of common rice (husked) was 20 seers 1 chittack, of wheat 19 seers 4 chittacks, and of barley 33 seers 5 chittacks per rupee. The contrast between prices at the present day and those of a century ago is even greater, as in December 1799 the price of the finest *arwā* rice was 32½ seers per rupee, while *sāthī* rice could be bought at the rate of 1 maund 5 seers per rupee. Even in 1803, a year of scarcity, the cheapest rice sold at over one maund per rupee, and the dearest at 20 seers per rupee; while barley was sold at 2¼ maunds for the rupee. The prices of grain have risen enormously during the last hundred years; but, on the other hand, there has been a very great growth in the income of all classes, and during the last generation the development of communications has had the effect of levelling prices over larger and larger areas. The loss of one or even two crops of the year has therefore a tendency to become less felt, as well as the effect of failures in isolated tracts. Besides this, the vast majority of labour is of an agricultural character, and is paid in kind; and immemorial custom has fixed the amount thereof, so that the high prices of grain affect a large section of the community less than would otherwise be the case. PRICES.

In Darbhangā the prices of common rice, *maruā* and maize are by far the most important in the effect they have on the material condition of the people. In normal years very little rice is consumed by the poorer classes except what they get while harvesting the crops, and *maruā* and maize constitute the poor man's food. During a famine year, however, the stocks of rice are the last to run out, and in the last famine first maize and then *maruā* almost disappeared from the market. The price of common rice is highest in August, and there is then a sharp fall in prices till December; after this there is a steady rise till August. The price of *maruā* is lowest in September, and it rises to its highest point in July, but there are intermediate falls in the price in December, April, and June, the two former being no doubt due to the fact that the winter rice and *rabi* are then coming into the market. Like *maruā*, maize is cheapest in September and dearest in July, but there are two intermediate cheap periods in February and April.

Famine prices.

In November 1873 Government declared that, whenever in any district the market price of common rice should rise to 10 seers for a rupee, famine rates should be considered prevalent in that district. In 1866 this abnormal rate was not reached in Darbhangā till the month of April; but in 1874, it prevailed from the commencement of the year. By April 1874 it had risen 20 per cent. higher, and thenceforward the quotations for rice were merely nominal, for none owned by private dealers was to be had in the market. The prevalence of higher prices in 1874 than those which prevailed at the corresponding period of 1866 was not restricted to the rice market only; it was also strikingly exemplified in the case of maize, which forms a staple article of food among the labouring classes. In January 1866 maize was selling at 18 seers for the rupee; in January 1874 the quotation was 13·9 seers; in April 1866 the price of this food-grain had risen to 14 seers; while in April 1874, when Government threw open its stores, the market price of maize was 12·9 seers for the rupee, that is, nearly three times the rate prevailing at that period in ordinary years.

In the scarcity of 1888-89 the price of maize never fell below 14 seers 13 chittacks and rarely below 17 seers; and in 1891-92 it never fell below 16½ seers. During the latter scarcity *maruā* only fell to the same price in August 1892, and common rice never sold below 12 seers. Such high prices only lasted a short time during this scarcity; but in the famine of 1896-97 the cheaper grains were practically out of the market in many parts of the district and for long periods, and common rice was the only food stuff available to the poorest labourers. The price of common rice was 9 seers per rupee and even less from December 1896 to August 1897, except for a short time in January; and it fell as low as 7 seers later. The price of maize was 10 seers or below from the middle of December to the middle of August, and for a time it was hardly procurable at all in some bazars from March to July 1897. *Maruā* was sold at 12 seers and below from the middle of August 1896, and from April to August 1897 the price oscillated between 9¼ seers and 12 seers. In the case of all these grains, the highest prices were reached in June and July, common rice being sold at 7 seers, maize at 8½ seers, and *maruā* at 9¼ seers per rupee. In 1906 the disastrous floods which swept away the crops, destroyed the roads, and interrupted traffic on the railway, sent up the prices to a higher level even than they reached in the famine of 1896-97. From August to November the price of *maruā* and maize varied from 10½ to 12 seers per rupee, while the average price of rice was 6½ seers, rising at one time as high as 5 seers per rupee.

MATERIAL CONDITION OF THE PEOPLE.

Special enquiries regarding the material condition of the people were made during the last settlement operations, and the result was to show that the average family of pure cultivators, who form over three-fifths of the population, should enjoy a moderate degree of comfort and be able to save sufficient in ordinary years to tide them over an occasional season of short crops. With regard to the labourers, the state of affairs is different. This class can only just make ends meet in an ordinary year, and has no reserves to fall back upon when the crops fail. A large part of it must therefore turn to Government for relief, when a serious crop failure produces a diminution of the demand for field labour, which is its chief means of support. The course of events in the famines of 1874 and 1897 shows, however, that the staying powers even of this class have increased considerably; and it is noticeable that a large proportion of the labourers were able to dispense with Government relief during the later famine. The improvement in the condition of the ryots was no less marked. Referring to 1873-74, and to the fact that the famine then was greatest in the district of Tirhut (which is now split up into Darbhangā and Muzaffarpur), Sir A. P. MacDonnell stated that the ryots of that district were so impoverished, so utterly without resource, and so unable to bear up against the failure of a single season's crop, that one-third of the population was at one period in receipt of relief from the Government. In 1896-97, when the distress was at its highest, more than three-fifths of the persons in the Division who were in receipt of relief belonged to the districts of Muzaffarpur and Darbhangā, but, instead of forming one-third of the population of those districts, they now formed less than 3½ per cent. of it. The opinion of the Government of Bengal on this point is contained in the following extract from the Resolution on the Commissioner's Report:—"That under less favourable circumstances than in 1873-74, the numbers requiring Government relief should have been so much less would appear to establish the fact of general improvement in the general circumstances of the people, but the whole conditions of the relief administration during the two famines were so different that the inference is not so conclusive as it would have been had the same methods been adopted in both cases. Nevertheless, the general experience and observations of the officers engaged in the famine, some of them with knowledge of facts, both then and now, as well as the concurrent opinions of non-officials acquainted with these provinces, do indicate that even in Bihār, during the past quarter of a century, there has been a considerable advance in material prosperity, and that the power of the country as a whole to withstand the calamities of seasons has greatly increased."

Indebtedness.

The statistics obtained during the last settlement show that only a little over 6 per cent. of the ryoti holdings and only $2\frac{1}{2}$ per cent. of the ryoti area are affected in whole or in part by sales or mortgages with possession. The statistics do not show cases in which the land is mortgaged as security for a loan but possession is retained by the borrower, nor do they disclose amounts borrowed on security other than land. But loans of the first kind are usually only made to persons whose credit is otherwise good, as the fear of a previous mortgage ordinarily deters lenders from advancing money on the security of the land without obtaining possession. As to loans of the second kind, the land is the ryot's chief source of credit, and only petty loans of a floating and temporary nature are made on other security. It is satisfactory, therefore, to find that so small a proportion of the ryots are indebted to such an extent that they have to part with their holdings or proportions of them in order to borrow money, and that the total amount of indebtedness incurred in this way is less than one-sixtieth of the share of the gross annual produce secured by the cultivating classes. The indebtedness of the Darbhangā agricultural community cannot, therefore, be said to be a very serious matter.

Regarding the whole question of the material condition of the people, Mr. Kerr summarises the position as follows: "Over four-fifths of the total population, or $2\frac{1}{3}$ million souls are entirely dependent on agriculture as a means of livelihood. Nearly half a million of these have either no lands at all, and are dependent on the wages of labour for a livelihood, or have holdings which are too small for their maintenance, so that they are compelled to supplement their incomes by working for other cultivators. More than half of this body of cultivating and landless labourers had to turn to Government for relief during the last famine, and their condition must always cause anxiety at times of short crops or high prices. The remainder of the agricultural community (omitting the small body of landlords) consists of the pure cultivators. These persons, numbering about $1\frac{3}{4}$ millions, enjoy a moderate condition of comfort. The average size of their holdings is over $3\frac{1}{2}$ acres, or more than an acre in excess of the area of the 'subsistence' holding. They can weather two or three successive bad seasons from their own resources without turning to Government for relief. Little more than a twentieth of them are in debt, to the extent that they have had to part with their holdings, or any portion of them, in order to borrow money, and the amount of their indebtedness is an insignificant proportion of the total annual income of the class to which they belong. The district requires for its own consumption less than three quarters of its total annual production of food grains in an ordinary year. The value of the surplus available for export is nearly double the annual rental of the district.

"Though the present condition of the pure cultivators is on the whole favourable, the district could not, under the present conditions of agriculture, support any considerable increase of population without a material reduction of the standard of comfort. Moreover, the dependence of a large part of the district on a single crop would render any further reduction in the average size of holdings a matter of grave apprehension. For the 'subsistence' holding is a holding from the produce of which an agricultural family can just support itself in ordinary years. When the subsistence holding comprises lands which are capable of growing a considerable variety of crops, it is exceedingly unlikely that all of them will fail in the same year. But where, as in a large part of Darbhangā, the holdings are composed of land only capable of producing winter rice, the failure of that crop must cause universal distress, if the holdings are at, or only slightly above, the subsistence limit. It is the predominance of the winter rice in Darbhangā which causes that district to suffer so severely from the effects of drought, and though the district might in an ordinary year support a slightly larger population than at present, a reduction in the average size of holdings to the subsistence limit might plunge the whole population into destitution at the next general failure of the rice crop, instead of only one-tenth of the population as at the last failure. But the last census figures indicate that no such rise of population, as would be necessary to produce this effect, is likely to occur, and it may be hoped that the population of the district will remain within such limits, as to enable the larger portion of it to continue, as at present, to enjoy a moderate degree of comfort, taking good years with bad."

CHAPTER VIII.

OCCUPATIONS, MANUFACTURES AND TRADE.

OCCUPATIONS. DARBHANGA is essentially an agricultural district, the vast majority of the inhabitants being engaged in agricultural pursuits. No less than 78·6 per cent. of the total population derive their livelihood from agriculture; and of these 46 per cent. are actual workers, including 12,000 rent-receivers and 758,000 rent-payers, while the landless labourers aggregate 262,000. Of the persons engaged in occupations other than agriculture, more than a sixth are found in the four municipalities, leaving only a little over half a million in the strictly rural area who are not dependent solely on agriculture; and of this half million nearly 5 per cent. follow agricultural pursuits to a certain extent. Only 10·2 per cent. of the population are supported by industries, 0·4 by commerce, and 1·1 per cent. by the various professions. Of the industrial population 57 per cent. are actual workers, among whom are 17,000 cow-keepers and milk-sellers, 16,000 fishermen and fish-dealers, 13,000 oil-pressers and sellers, 14,000 vegetable and fruit-sellers, 8,000 grocers, 9,000 cotton-weavers, 7,000 potters and 8,000 basket and mat-makers, besides numerous toddy-sellers, tailors, gold and silversmiths, ironsmiths, carpenters and shoe-makers. Of the professional classes 48 per cent. are actual workers, including 5,000 priests, 4,000 religious mendicants and 800 teachers. Among those engaged in other occupations are 13,000 herdsmen, 7,000 beggars and 63,000 general labourers.

MANUFACTURES. As might be expected in a district containing no important trade centres or large manufacturing towns, the industrial population is very small; and the bulk of the artisans are engaged in supplying the simple needs of a rural people. This economic backwardness is the result of the dependence of the majority of the people on agriculture. They live in villages dotted among their fields, and each village constitutes, to a large extent, a self-sufficing industrial unit. The villagers grow their own food, grind their own corn, and build their own houses. In spite of the introduction of manufactured piece-goods, their thread is still, to a certain extent, spun by the women and woven by the village weaver, while the rough implements needed in agriculture are made and repaired by the village smith and carpenter. Like the adjoining districts, Darbhangā contains a smaller industrial population than is to be found in the South Gangetic districts of the Patna Division. Native handicrafts have not found the same home in North Bihār as in the districts to the south of the Ganges; and it has been suggested that the reason for this is that, after the murder of Alamgīr and the fall of Delhi in 1759, many members of the Muhammadan nobility retired to their *jāgīrs* in Patna, Gayā and Shāhābād, bringing with them large numbers of artificers, while the districts to the north of the Ganges were still in an unsettled state, sparsely populated and only partially cultivated. However that may be, it is certain that in Darbhangā with the exception of the industries that have been developed by European enterprise, manufactures in the proper sense of the word are few in number and of little importance. By far the most important industry is the manufacture of indigo, which is entirely in European hands; and the only other large industries are cloth weaving, which is now in a languishing condition, and the manufacture of sugar, salt and saltpetre, all of which were originally fostered by Europeans. A description of the indigo industry will be found in the next chapter, and a brief account of the other manufactures is given below.

Saltpetre. In Bihār, which is the chief source of saltpetre in India, the conditions for the natural production of the compound closely approach the theoretical ideal. With a population exceeding 500 per square mile, where agriculture is the chief occupation, and where there is consequently a high proportion of domestic animals, the soils round the villages have an abundant supply of organic nitrogen. The climatic conditions of temperature and humidity are also unusually favourable for the growth of so-called nitrifying bacteria, which convert ammonia by successive stages into nitrous and nitric acid. Wood and cow-dung are largely used for fuel, and the immediate vicinity of each village thus forms a perfect laboratory for the formation of potassium nitrate. In the long period of continuous surface desiccation, which follows a small monsoon rainfall, the compounds so formed in the soil are brought to the surface by capillary action, and appear as a white efflorescence of dried salts, which is collected and purified for export as saltpetre.

Before the discovery of large deposits of sodium nitrate in Chili, India had almost a monopoly of the supply of natural saltpetre, upon which Europe largely depended for the manufacture of gunpowder: and the greater part of this supply came from Bihār. The production of saltpetre was of especial importance during the long wars with France, and the great fluctuations in its price gave rise to heavy speculations during periods of international complications. A system which provided for the control of manufacture and sale of salt by the agency of the Company's servants was accordingly introduced by Clive and Warren Hastings in 1765-80; and the production of saltpetre in Tirhut long continued to be supervised by Europeans. As late as 1847 there were four factories under European supervision in Tirhut; but the fall of prices caused the Europeans who were engaged in the trade to withdraw their capital, and the manufacture is now entirely in the hands of natives. Of late years a series of bad seasons, combined with low prices in Calcutta, has had an injurious effect on the manufacture, and many refineries have been closed. The outturn of saltpetre has accordingly fallen gradually from 64,700 maunds in 1895-96 to 38,000 maunds in 1904-05; though the outturn of salt deduced during the process of manufacture has remained fairly constant, and has risen from 2,760 to 2,910 maunds.

The manufacture is in the hands of a poor and hardy caste called Nuniās, and is founded on a system of advances made to them by middlemen, who again contract with the larger houses of business in Calcutta. It is controlled by the Northern India Salt Department, which grants licenses for refining salt, for making saltpetre and for the manufacture of the unrefined saltpetre called *khārī*. The process of manufacture is simple, and the implements employed are very primitive. Manufacture is carried on in small factories, situated at towns and villages scattered over the country. Nitrous soil is collected from the vicinity of habitations and is lixiviated in earthen filters, after which the nitrous brine is concentrated in small iron or earthen vessels with the aid of artificial heat, and saltpetre is obtained by crystallization as the temperature of the concentrated liquid falls. The saltpetre so obtained is impure in quality, as it contains earthy matter and foreign salts (such as chloride of sodium) in mechanical admixture. This impure saltpetre is collected in refineries situated at different points among the village works, and after being purified in them to a fair degree of refraction, is sent to Calcutta, where some of it is purified to a higher degree of refraction, and some is exported to the United Kingdom, the United States, China and other countries.

Sugar manufacture.

The manufacture of sugar has long been carried on to a considerable extent in Darbhangā, where many of the European indigo concerns were originally started as sugar factories. The manufacture of sugar was, however, given up when indigo proved more profitable, and for the last half century the industry has been almost entirely in native hands. For some years past the manufacture has been practically confined to the Madhubanī subdivision, where it does not appear to have suffered materially from the competition of cheap imported sugar. There were 32 refineries at work in 1895-96 with an outturn of 43,000 maunds valued at Rs. 4,30,000. Since then two refineries have been closed, and in 1904-05 the outturn from the 30 factories still at work was 41,400 maunds, valued at Rs. 2,71,250. The outturn has, therefore, not decreased appreciably of late years, though the fall in prices has led to a great reduction in the value of the product.

Cotton fabrics.

Darbhangā is one of the few districts in Bengal in which cotton weaving is still comparatively an important industry. It is the chief centre in Bihār of the manufacture of the coarse cloth called *khokti*, which is woven in the Madhubanī subdivision. This cloth derives its name from the species of cotton of which it is made, *khoktibanga* cotton (*Gossypium herbaceum*), which is indigenous to the subdivision. The cloth turned out is naturally of a brown colour, resembling tusser silk both in colour and texture; and its value varies from Rs. 5 to Rs. 100 for a piece of 40 yards. It is said that the best quality equals, if it does not surpass, good brown Holland, and that it can well be worn as a summer dress. The cloth is very durable, and is smooth, glossy and of a fine texture. The manufacture is now languishing for want of patronage and owing to the competition of European piece-goods.

Besides *khokti*, a coarse cotton cloth called *motiā* or *gāzi* is woven. This is a coarse cloth, which is chiefly used by the poorer classes in the cold weather. It is warm and durable, and is used for making *dhotīs*, jackets, wrappers and quilts for men, and for *sārīs* and bodices for women. The cloth is popular among the poorer classes, who cannot afford to purchase woollen fabrics for the winter. There is still some demand for it all over Bihār, and the industry, though not very extensive, is said to be flourishing. Attempts have been made to give a stimulus to the manufacture by the introduction of the fly-shuttle loom, but so far with but little success. Repeated efforts during the last few years to popularize the use of the fly-shuttles among the local weavers have failed to arouse even a passing interest; and all trouble and expenditure in that direction have proved useless. The difficulty as to the high price of the improved fly-shuttle looms has been overcome by adapting the fly-shuttle to the ordinary country looms. The outturn obtained from these modified looms is said to be almost equal to that of the framed fly-shuttle looms, *i.e.*, nearly three times that of the ordinary looms in vogue in the district, while their cost scarcely exceeds that of the country looms. Trained weavers have been sent out with their looms to various busy centres in the interior to teach the local weavers their use; and weaving classes have been opened at Laheriā Sarai, at which both the fly-shuttle loom of the Serampore pattern and the modified loom are worked. The classes are, however, very thinly attended, and but little progress has been made, owing to the conservatism of the weavers.

Other industries. The other industries carried on in the district are of little importance, and consist mainly of small village industries, such as the manufacture of pottery, mats, baskets, etc. The one indigenous industry which has not suffered from foreign competition is the manufacture of brass; but the workmen are rarely very skilful, and the articles produced are of poor design, with the exception of the utensils manufactured at Jhanjhārpur, a village about 14 miles south-east of Madhubanī, which have a local reputation. A small quantity of blankets are made for local use by the shepherd caste, the Gareris; and some thousands of pairs of shoes are manufactured yearly by the local cobblers, but the exportation of skins and the consequent rise in the price of leather have injured the trade. Lac bangles are made by Laheris and are exported to other districts.

Factories. There is only one factory in the district, the locomotive workshop under the management of the Bengal and North-Western Railway at Samāstipur, which employs an average of about 1,000 hands on the repair of railway engines, carriages and waggons.

TRADE. The principal exports are rice, indigo, gram, pulses, linseed, mustard seed, saltpetre, tobacco, hides, *ghi* and timber; and the imports are rice and other food-grains, salt, kerosene oil, gunny bags, coal and coke, European cotton piece-goods and raw cotton. Among exports rice, as a rule, occupies the most prominent position, closely followed by linseed and unmanufactured tobacco. Most of the rice is exported to the United Provinces and the neighbouring districts of Sāran, Muzaffarpur, Monghyr and Bhāgalpur; nearly the whole of the linseed is sent to Calcutta; while the largest importers of the local tobacco are the United Provinces and the other districts of Bihār. The only other articles exported in large quantities are gram and pulse, other food grains and unrefined sugar. Calcutta and Burdwān take the largest proportion of gram and pulse; the food grains are nearly all sent to adjoining districts; and a large proportion of the sugar finds it way to the districts of the Bhāgalpur Division. The most important imports are rice, paddy and other food grains, which come for the most part from Bhāgalpur and Nepāl; coal and coke, which are imported from Mānbhūm, Hazāribāgh and Burdwān; salt, which is brought in from Calcutta; kerosene oil, which is imported from the oil godowns in the 24-Parganas; and European piece-goods from Calcutta.

Most of the trade with the neighbouring districts is rail-borne; but there is a considerable volume of trade between Darbhangā and Nepāl, which is carried on by means of carts and pack-bullocks, and to a certain extent by means of coolies. The imports from Nepāl consist chiefly of grains, such as rice, paddy, gram, pulse and oil-seeds, and of timber floated down the rivers. The most important exports to that State are European and Indian piece-goods, salt and kerosene oil. Darbhangā accounts for 14·8 per cent. of the total trade of the Patna Division, but its commercial life is far from vigorous, and the last quinquennium shows a falling off of 4 per cent. in its trade.

Trade centres. The principal centres of trade are Darbhangā and the outlying subdivisional head-quarters, Madhubanī and Samāstipur. Roserā, owing to its position on the Gandak, was at one time the largest market in the south of the district, but it has lost much of its former importance since the opening of the railway, though it is still a flourishing bazar. In the north, Narahiā in the Phulparās thāna is an important centre for the Nepalese grain traffic; and there are a number of big bazars in the same thāna, where the Nepalese trade changes hands. The other markets in the district are mainly of local importance; the largest are Pūsa, Kamtaul, Dalsingh Sarai and Jhanjhārpur.

Fairs. A considerable amount of trade is also carried on at the annual fairs held in various parts of the district, the most important of which are the Kusheswar Asthān, held at Reota, 38 miles south-east of Darbhangā, on the 14th Phāgun (February-March), and the Sivarātri fairs held at Saurāth and at Kapileswar, 4 and 7 miles respectively from Madhubanī. Other large fairs are held at Silānāth and Dubhī, 15 miles from Khajaulī, at Pīpra Ghāt, 30 miles south-east of Darbhangā, at Sultānpur in the jurisdiction of the Mohiuddīnnagar outpost, and at Mahādeonāth in the Phulparās thāna. These are all primarily religious gatherings; but they are attended by large numbers, and a brisk trade is driven at them by traders, who supply the crowds with various articles of commerce. Here, as elsewhere in the district, the chief traders are Agarwāls, Barnawārs, Kasarwānis, Kathbaniās, Khattrīs and Sinduriās.

Weights and measures.

There is no uniform system of weights and measures in the district. Both *kachhā* and *pakkā* seers are recognized; the standard seer of 80 *tolās* is generally used in the towns, but even in Darbhangā town a local seer of 88 *tolās* is also common; while a seer of 50 or 52 *tolās* is used in the villages. For gold, silver and country medicines the following weights are used:—1 *tolā*=12 *māshās* and 1 *māshā*=8 *ratis*. There is a similar diversity of practice with regard to measures of length. Cloth dealers use one yard of 36 inches and another of 40½ inches; and tailors a yard of 41 inches. For measuring lands the pole or *laggī* is usually employed, and this is equivalent to 9 feet 8½ inches. The other measure of length in common use is the *hāth*, which is usually equal to 18 inches.

The measures of capacity generally recognized by the people are the *jhabbā*, *chungā* and *pailī*. The *jhabbā* is used for milk and oil, and the *chungā* or *māpa* for wine, oil and other liquids; both are regarded as equal to a seer of 80 *tolās*. Another measure, also used for oil and milk, is called *pailī*, if made of bamboo, or *nāpna*, if made of earthenware: this is equal to a seer of 52 *tolās*.

CHAPTER IX.

THE INDIGO INDUSTRY.

PROGRESS OF THE INDUSTRY.

INDIGO was a product of North Bihār long before the advent of the British, but its cultivation by European methods appears to have been started by Francois Grand, the first Collector of Tirhut. Writing in 1785 three years after his appointment as Collector, he claims to have been the pioneer of the industry, and says:—"I introduced the manufacturing of indigo after the European manner, encouraged the establishment of indigo works and plantations, and erected three at my own expense." It is at least from this time that the manufacture of indigo began to develop into an industry, and to attract European enterprise; and by the end of the 18th century the present concerns of Dalsingh Saria, Jitwārpur, Tiwāra and Kamtaul had been founded. In 1788 there were 5 Europeans in possession of indigo works in Tirhut; in 1793 the number of factories had increased to nine; and by 1803 altogether 25 factories had been established. During these early days, the industry was directly fostered by the East India Company, and special permission had to be obtained by Europeans wishing to engage in it. In 1802, however, the Board of Directors passed orders that no further advances or other pecuniary encouragement should be given to the planters, as the large profits obtained from the sale of the product made such aid unnecessary. Indigo accordingly became an independent and self-supporting industry, the pioneer planting industry in Bengal.

Its progress during the next few years was rapid, though there appear to have been many failures, probably owing to over-production. In a report submitted in 1810 the Collector of Tirhut stated that, taking one year with another, the district seldom sent less than 10,000 maunds of indigo to Calcutta for export to Europe, that 30,000 to 50,000 souls received their principal support from the factories, and that on the average each factory disbursed from Rs. 25,000 to Rs. 30,000 per annum in hard cash to the labourers and cultivators for some miles round. He estimated that in this way not less than six or seven lakhs of rupees were circulated every year by the planters in Tirhut, and urged that the advantages of the industry to the labouring classes were so great that Government should encourage it in every possible way. "Let the speculator win or lose," he wrote, "acquire a princely fortune or die a pauper, the district is equally benefited by his industry, and his struggles for prosperity do rarely succeed. Some of the planters succeed, but the majority of them fail." Difficulties appear to have arisen later through the competition of rival concerns, and in 1828 the Collector represented that indigo cultivation had extended so greatly that some restriction on it was desirable for the benefit of the district. "From the misunderstanding," he wrote, "which has prevailed and still prevails amongst the European planters, disputes with one another are of very frequent occurrence: disputes have, however, of late occurred through descendants of Europeans embarking in indigo cultivation, chiefly, if not entirely, by native agency. For the peace of the district and welfare of the established planters, it therefore appears highly desirable that the Government restrictions regarding the erection of factories by Europeans should be extended to the descendants of Europeans, and power be vested in the Magistrate to prevent engagements for the cultivation of indigo plant by other than the proprietor or proprietors of one established factory."

In 1850 there were no less than 86 factories in Tirhut, several of which were used for the manufacture of sugar, but about this time sugar was finally superseded by indigo as the European industry of the district, and many refineries were converted into indigo concerns. By 1874 there were altogether 126 factories and outworks engaged in the production of indigo, and the area under cultivation was nearly 100,000 acres. At this time, Darbhangā contained the largest concern in India, Pandaul, which with its outworks comprised an area of 300 square miles; it was subsequently split up, the northern outworks being purchased by the Mahārājā of Darbhangā, who abandoned the cultivation of indigo in them a few years ago. Difficulties were now threatened by the feeling of tension between the ryots and the factories produced by certain abuses which had crept into the system of cultivation. A report submitted to Government by the Commissioner of Patna in 1877 showed that the system prevailing involved an amount of lawlessness and oppression, principally in the shape of extorted agreements to cultivate and of seizure of ploughs and cattle, which could not be tolerated. On receipt of this report, some of the leading planters as well as the officials of Bihār were consulted through the Commissioner.

It was an object to do nothing which would unduly excite the mind of the ryots, and to avoid any such agitation as might lead to breaches of contract and general embitterment of relations between planters and ryots; and as some of the leading planters declared themselves sensible of the necessity of reform and willing to assist in the work, and for this purpose undertook the establishment of a Planters' Association, any action on the part of Government was postponed, and the matter was entrusted to their hands. This body showed a sincere desire to place the relations between planters and ryots on a more satisfactory footing, and drew up a series of rules, embodying very important reforms, for the guidance of its members. Owing to the efforts of the Association, there has been no recurrence of the old complaints, and very cordial relations have existed up to the present between the planters and the cultivators.

Until the discovery of the Badische artificial dye the area under cultivation appears to have been steadily on the increase, and by the end of the 19th century indigo had spread into every thāna of the district, though it was always more prevalent in the north, where the soil has never been altogether suitable for the crop. The industry is now suffering from the competition of the artificial dye in Europe, and from the high prices of food-grains and the consequent demand for land in Bihār. The price of the natural dye has consequently fallen, and many factories have had to abandon or contract very greatly the growth of indigo. At the time of the last settlement 52,136 acres or 3 per cent. of the cultivated area were under indigo, while there were 28 head factories, with 36 outworks in the district. In 1904 their number had fallen to 24 with 27 outworks, and the area under cultivation had also diminished considerably; it was estimated at 34,000 acres in 1903-04, while the final forecast of the indigo crop in Bengal returned the area sown in 1906 as 28,400 acres. Government has come to the aid of the planters with substantial grants for scientific research, the aim of which is to ascertain whether it is possible to increase the outturn and quality of the dye at a cheaper cost; excellent work in the chemistry, bacteriology and agriculture of indigo has been done and is still progressing; and every effort is being made to improve the quality of the plant by importing fresh seed from Natal. But so far these experiments have not succeeded in arresting the decay of the industry. The price obtained for indigo is barely sufficient to cover the cost of production, and many factories are either closing altogether or are reducing the area cultivated with indigo, growing in its place sugar, cotton and other country crops. Most of the area now under indigo is in the Sámāstipur subdivision; the plant is grown to a smaller extent in the head-quarters subdivision; and the industry is no longer of any practical importance in Madhubanī, as all the factories in the north and east have ceased to grow the crop.

The gradual decline of the industry in recent years will be sufficiently apparent from the following statistics showing the outturn, value and price of the indigo manufactured:—

Year.	Outturn in maunds.	Value in Rs.	Price per maund in Rs.
1895-1900 (average) ...	11,599	21,34,895	184
1900-01	9,540	12,87,900	135
1904-05	1,673	2,50,950	150

Cultivation. The land on which indigo is to be grown is prepared for sowing as soon as the *kharīf* crops have been reaped, as it is of great importance that the soil should retain the moisture supplied by the rainfall in October and November. The land is ploughed and reploughed until the clods are all pulverized, and after being manured, it is levelled and smoothed with a plank roller composed of a long heavy beam on which two men stand. The seed is sown at the beginning of the hot weather, as soon as the nights begin to get warm, a special drill, with coulters about 5 or 6 inches apart, being used for the purpose; and after sowing, the roller is again used to level the surface. The seedlings are very delicate until their roots are well developed, and many perish owing to dry west winds; but moist east winds after sowing, and spring showers later, are very beneficial to the young plants. They make slow progress until the monsoon sets in, when the growth becomes very rapid; and they are ready for cutting, which takes place immediately before they flower, in July or August. A second crop is obtainable in September, but usually yields less than the first crop, the outturn of which is ordinarily 80 to 120 maunds of green plant per acre. The yield of 100 maunds of good ordinary plant should be about 10 seers of indigo.

Soils and manures. Indigo may follow indigo, but is more generally rotated with such crops as sugarcane, tobacco, poppy, cereals and oil-seeds, as it is an exhausting crop, which cannot well be grown on the same land for more than three successive seasons; on the other hand, being a deep-root crop it forms an excellent rotation crop for those which have surface roots, as is the case with many food-grains. It is usually grown on high lands beyond the reach of floods, the soils are varied in character and composition, but deep alluvium loams seem to suit the crop best. Many soils of this description are deficient in phosphoric acid and nitrogen, but are generally rich in other useful constituents; while extensive experiments have proved that superphosphate and nitrate of potash can be economically applied. The refuse indigo plant (*sīth*) is the manure most easily obtained, and is very valuable; but it is less suited for indigo itself than for rotation crops, such as sugar-cane, tobacco, poppy, cereals and oil-seeds. It produces heavy crops of indigo, but the leaf is deficient in colouring matter; and indigo grown on land heavily treated with *sīth* is liable to injury from insect pests. Farm manure, chemical manures, such as saltpetre and lime, bone-dust and oil-cake are also used.

The seed used in Tirhut comes for the most part from the United Provinces, as there is a general belief among the planters that the best seed is obtainable there and that local seed does not keep good from season to season and does not germinate properly. The system of getting seed in this way, without any special selection, has however caused deterioration in the varieties commonly grown, and there is little doubt that the plant commonly cultivated does not now produce a satisfactory amount of dye matter, particularly on worn-out indigo lands. The chief cultivated form is not *Indigofera tinctoria*, as was formerly supposed, but *Indigofera sumatrana*, which was introduced about 150 years ago. Seed.

Within recent years Natal indigo (*Indigofera arrecta*) has been introduced, the seed being obtained direct from Natal and also from plants acclimatized in Java. This plant has been found to give a very considerable increase of colouring matter from the unit area of land, and will produce excellent cuttings for two years in succession and mediocre plants for a third year, whereas other varieties have to be resown annually. It has a much more vigorous habit of growth than the old variety, and the leaf contains a larger proportion of the colour-yielding principle. It appears to be eminently suited to the soils and climate of Bihār, and farms have now been established in three districts for the cultivation of its seed in an extensive scale.

The colouring matter from which indigotin is derived exists almost entirely in the leaf of the plant. It increases as the plant grows, but deteriorates after a certain stage, and harvesting and steeping have therefore to be carried on expeditiously. Plants which have been cut some time and become blackened by heating in bulk contain very little dye matter, so that the green plant cannot be carted very far. A plant which is forced by manure to very active growth also gives a poor percentage of dye matter. Colouring matter.

MANUFACTURE. After they have been cut the leaves are taken to the factory and are there steeped in large vats until fermentation is complete. The old system of treating the plant requires two sets of vats, one on a lower level than the other, those on the highest level **Steeping.** being used for steeping the plant, which is kept submerged by logs of wood or bars fixed in position. During this process active fermentation takes place through the action of soluble ferments (enzymes), and causes the formation of a compound which is easily convertible into indigotin by the action of air. The period of steeping varies with the temperature of the air and water; if the temperature of the water is 90° to 92° F., steeping for 10 hours is sufficient, but instead of varying the time, it is preferable to heat the water in the reservoir to a definite temperature. It has been shown by experiment that when the plant is steeped in water at 150° to 160° F., the colouring principle is extracted in half an hour; and indigo made in this way is superior in quality and contains about 75 per cent. of indigotin.

Oxidation or beating process. When fermentation is complete, the liquid in the steeping vats, which varies in colour from bright orange to olive green, is drained off into the lower vats, and is there subjected to a brisk beating, the effect of which is to cause oxidation and separate the particles of dye. As the oxidation proceeds, dark blue particles of indigotin appear in the liquid, the colour of which consequently changes, and the beating is continued until a little of the liquid placed in a saucer readily throws a dark blue precipitate, itself remaining of a clear amber colour. If there is any delay in oxidation, there is a considerable loss of colouring matter, and the indigo produced is inferior. Oxidation was at one time accomplished by hand-beating, but in most Bihār factories it is now done by a beating wheel worked by power from a central engine.

Lime and ammonia process. The improved method of treating the plant known as Coventry's lime and acid process, which is used in a few Bihār factories, requires a vat intermediate between the steeping and beating vat. Lime is added to the indigo liquor, and a precipitate of calcium and magnesium carbonates then forms, which also carries down various other impurities. The cleared liquor, when run off into a lower vat and oxidized, yields indigo of good quality, and a substantial increase of colouring matter is obtained. An ammonia gas process patented by Mr. Rawson in 1901 also produces a direct increase of colouring matter.

Boiling and final preparation. Finally, the sediment (*māl*) which remains in the vat is boiled, strained and made up into cakes for the market. The first process in these final stages of manufacture is to boil the precipitate which settles after oxidation; the indigo produced from it is improved, if sulphuric acid is added. The dye matter is next placed on a cloth strainer until it becomes fairly dry. It is then carried to the press and subjected to gradually increasing pressure until it has taken the form of firm slabs, which are cut into cakes and slowly dried on racks. Good indigo should contain 60 per cent. or more of indigotin, should be bright and of a dark blue colour, with a coppery gloss, and should break with an evenly coloured fracture.

LANDED INTERESTS. The chief feature of the industry in this district as compared with the other indigo-growing tracts in North Bihār is the large area cultivated direct by the factories themselves: it was, in fact, ascertained in the course of the last settlement operations that the factories in the Samāstipur subdivision had in their direct cultivation no less than 94 per cent. of the total area under indigo. The area held by them as landlords is far smaller than elsewhere in North Bihār, amounting to only 6 per cent. of the total area of the district. The fact that the Darbhangā factories grow the greater part of the crop themselves, instead of merely purchasing it from others, has been of great advantage to them in the present depressed state of the industry, when the falling price of the natural dye has made the ryots unwilling to grow a crop which does not pay them so well as ordinary crops.

The total area held by them as proprietors or permanent tenure-holders is even smaller, being under 1 per cent. of the total area; and the greater part of their interests as landlords are derived from temporary tenures. The reason for this is that a factory has seldom an opportunity of buying an estate with lands situated conveniently for its purpose. The sale of estates is regarded as a social disgrace only to be resorted to in the last extremity, and consequently proprietors will not part with their rights unless absolutely forced to do so; while the practice of granting permanent leases has almost entirely died out with the rise in the value of land. Factories are, therefore, mainly dependent on temporary leases for acquiring interests in villages in which they wish to extend or maintain the cultivation of indigo. Such leases are granted as security for loans or are simple farming leases (*thīkā*). The latter are due to the financial embarrassment of proprietors and to their desire to avoid the troubles of management. The term of the lease may vary from 5 to 20 years, and its renewal is generally made an opportunity for increasing the rent.

The *thīkā* leases are the commonest of all; and it is to the *thīkā* system and to his influence as a considerate landlord that the planter owes the strength of his position. The other class of leases common in the district consists of usufructuary mortgages, under which the factory grants a loan at a moderate rate of interest and receives the land of an embarrassed proprietor as security. Leases of this kind are either *zarpeshgī* or *sadua-patua.* In the former case the interest on the loan is paid yearly by deducting it from the rent payable to the mortgagor, and the principal is repayable on the expiry of the lease; in the latter both principal and interest are liquidated by deduction from the yearly rent due to the proprietor, and the tenure thus returns to the latter free of encumbrance at the end of the term agreed upon. The *zarpeshgī* leases are most in favour with the factories, as the proprietor is frequently unable to repay the principal on the expiry of the lease, and the factory consequently acquires a quasi-permanent interest in the land.

Under-tenures.

In some cases, factories take a lease of an under-tenure, this lease being known as *katkanā*, *e.g.*, if two factories quarrel about their respective jurisdictions, a sub-lease from one to the other generally forms the basis of a compromise. Again, a proprietor is prepared to grant a lease of his estate to a factory on condition that it takes the whole, but part may fall within the jurisdiction of another factory. In such a case, the good services of the Indigo Planters' Association are called in to arrange for the latter factory taking a sub-lease from the former, and thus the danger of friction is avoided.

The *tinkathiā* system.

A factory, taking a lease of an estate, acquires direct possession over all the lands which were formerly cultivated by the proprietor, and also over any lands which may become vacant during the period of the lease, by abandonment or surrender on the part of their original occupants. In addition to this, it was formerly the universal custom for the ryots to surrender to the factory for indigo cultivation a certain proportion of the lands of their holdings, usually three *kathās* in the *bīgha*, the ryots receiving an abatement of rent for the area so given up, and a promise that it would be returned to them on the cessation of the factory's connection with the village. This system of acquiring lands was always intensely unpopular with the ryots, and is not now generally practised by the best factories. As indigo cultivation usually entails the breaking down of field boundaries and the amalgamation of many small plots into one large one, it was generally practically impossible for any factory, even with the best will in the world, to trace out and restore to the ryot his original plot, on the expiry of a long-term lease; and consequently the system was a fruitful source of dispute and discontent.

It is satisfactory, therefore, that it is being abandoned, and that most of the land, which is found in the possession of a factory in its capacity as a temporary tenure-holder, is that which it has acquired unobjectionably as the representative of the superior proprietor.

Ryoti interests. The very considerable area held by the factories on ryoti interests is mainly acquired by the purchase of ryoti rights, either privately or through the Civil Court. The law facilitates the acquisition of occupancy rights by tenure-holders in an area like Samāstipur, where petty proprietors predominate and great subdivision of proprietary rights prevails. A person who does not hold the whole of the proprietary interests of an estate or *pattī* in farm, is not debarred from acquiring occupancy rights in the lands of the estate during the term of his lease; and consequently a factory, holding only a share of an estate in lease, as is very common, is frequently able, on the expiry of the lease, to retain in its own possession lands in which it has managed to acquire an occupancy right during the continuance of the lease, by means of direct cultivation and payment of rent to the non-leasing proprietors. This method of acquiring occupancy rights is of considerable importance in Darbhangā, for the factories in Samāstipur have occupancy rights in nearly a quarter of the land in their own cultivation.

***Kurtaulī* leases.** The land held by the factories as under-ryots is mainly acquired by what are known as *kurtaulī* leases, which correspond to the *sadua-patua* leases granted by proprietors to tenure-holders. The factory gives the ryot an advance of so many years rental of the land taken up, and in return is allowed to cultivate the land for that period, giving it back to the ryot on its expiry. The only risk run by the factory is that the ryot may go off with the advance, without paying the rent to the superior landlord, who may then sell up the holding and refuse to recognize the factory in any way. But in practice this is not a serious danger, for *kurtaulī* leases are generally executed for part holdings only, and the ryot remains in the village to cultivate the portion which he has not sublet to the factory. An analogous form of mortgage sublease is the *sud-barna*, in which, as in a *zarpeshgī* tenure, the factory gives an advance, on which the interest only is liquidated by deduction from the annual rent for the land sub-let, the factory retaining possession until the principal is repaid. But, here again, want of security other than the land, which is worthless if the ryot should abscond, prevents the system from being very common.

SYSTEMS OF CULTIVATION. The three main systems of indigo cultivation commonly practised are *zirāat* or direct cultivation by means of hired servants, *āsāmiwār* or cultivation through factory tenants, and *khushkī* or cultivation through outside ryots.

Zirāat. The term *zirāat* includes all land in the direct occupation of the factory, whether held by it as proprietor, tenure-holder, ryot or under-ryot. In Darbhangā by far the greater part of the area under indigo is cultivated direct, and it is estimated that the amount of indigo not grown direct by the factories cannot exceed 10 per cent. of the total.

Asāmīwār. When the system of *āsāmīwār* cultivation is followed, the indigo is grown by the factory tenants at fixed rates per *bīgha*. Generally documents, called *sattas*, are executed, the ryot usually receiving an advance and binding himself to grow indigo on a certain specified portion of his holding, and to pay damages if he should fail to carry out his agreement. All the expenses of cultivation are paid by the ryot, but the seed is given by the factory, which also cuts and carts away the indigo, the ryot being paid for the indigo at a rate fixed by the Indigo Planters' Association.

Khushkī. Agreements executed by ryots who are not the tenants of the factory are called *khushkī sattas* or voluntary agreements. In this case, the factory merely supplies the seed and pays for the crop when delivered ; it sometimes also gives an advance to the cultivator at a light rate of interest. The amount of *khushkī* cultivation in Darbhangā is very small, as indigo, if it is to pay, requires selected lands, carefully cultivated and rotated in an intelligent manner. These conditions are all wanting in the *khushkī* system ; the rate of remuneration has to be high in order to induce the outside ryot to grow indigo ; and the factory cannot therefore afford this system of cultivation.

INFLUENCE OF THE INDUSTRY. Regarding the general effect of the industry on the district, the following opinion of the Settlement Officer may be quoted :— "The ordinary cultivator in Darbhangā is little affected by indigo cultivation, except in so far as he may have an indigo factory for his landlord ; and as this usually implies protection from enhancement of rent, it is a pure gain to him, provided the factory does not force him to grow indigo against his will, or to give up his lands for the cultivation of indigo. The small area in which indigo is grown, otherwise than by the factories direct, renders the first danger inconsiderable, and the second has been minimized since the practical abandonment of the *tinkathiā* system. Hence it may be said that the cultivators of the district derive nothing but advantage from having indigo factories as their landlords ; and how great the advantage of stability of rents is, can only be appreciated by those who have seen and realized the constant and vexatious enhancements which, prior to our operations, were always going on in the estates of ordinary native landlords.

"The indigo industry may, without any qualification whatever, be pronounced a boon to the proprietors and labourers of the district. The manner in which embarrassed proprietors turn to indigo factories for loans has already been described. Even for proprietors who are not financially embarrassed, it is no small benefit to be able to rid themselves of the trouble of management, by handing over their property to a tenure-holder, who will pay a full rent punctually and manage the estate efficiently. The benefit of the industry to the labourers is clear. It has been calculated that on the average 172 labourers are required for one day per annum per acre for the cultivation and manufacture of indigo. This means an expenditure on wages alone of at least Rs. 20 an acre, so that the annual total wages bill for the 50,000 acres under indigo* must exceed ten lakhs of rupees. The cultivation of none of the ordinary food crops gives employment to so large a labour force as this, and the benefit of the indigo industry to the labourers of this district is enhanced by the fact that a good deal of the work has to be done in the hot weather when little other employment is available.

"It may be said, therefore, that the indigo industry is an unquestionable benefit to all classes in the district, with a possible reservation as regards the ryots who grow the crop on the *satta* system, and whose profits are not so large as they would derive from the cultivation of other crops. They are, however, compensated in other ways, and, in any case, their number in Darbhangā is so small as to render them of little importance in estimating the effect of the industry on the district as a whole, and in this connection the valuable services rendered on many occasions by members of the planting community to the general administration should not be forgotten."

To this it may be added that the planters have consistently shown themselves true friends to the cultivators and labourers in periods of adversity. Their readiness to help the latter was very clearly shown in the last famine of 1896-97, and the value of their services at this time of distress may be gathered from the remarks of the Commissioner, who wrote:—"The planting community, as in 1873-74, proved to be of inestimable value in the crisis. In the former year many of them were stimulated by the prospects of pecuniary advantage: in 1896-97 no such stimulus was offered; but at an early stage of the operations their services were offered gratuitously—an offer which

* The area under cultivation has since decreased to 28,400 acres.

they more than redeemed. Numbers of them sacrificed time, ease and health to assist Government, and many of them have been losers by their public-spirited efforts. Yet the work has been cheerfully done, and the community have once more proved themselves invaluable to the administration."

PRINCIPAL FACTORIES.

The following are the principal indigo factories in the district :—

FACTORY.	Outworks.	Factory.	Outworks.
HEAD-QUARTERS SUBDIVISION.		SAMASTIPUR SUBDIVISION.	
Anar.		Banhār.	
Buchauli.		Chal Mehsi.	
Baghauni.		Gangauli ...	Alampur.
Benīpur ...	Harsinghpur.	Harsinghpur	Balampur. Bhawāra. Rahīmābād.
Daulatpur ...	Meghaul. Rāmnagar.	Ilmāsnagar ...	Masena.
Hathauri ...	Rasulpur.	Jitwārpur ...	Dāūdpur. Hasauli.
Hāthi.		Keota ...	Chatrā. Gobindpur. Kamlā. Pembarhanda. Shāhpur. Tappa.
Mangalgarh	Kursauli. Motīpur.	Khāu Mirzāpur.	
MADHUBANI SUBDIVISION.		Maniārpur ...	Budaya. Kalyānpur.
Jaynagar ...	Narārh.	Muktapur ...	Kalyānpur.
Pandaul ...	Lahrā. Lohat.	Rewari.	
Rayām.		Shāhpur ...	Undi. Barauli.* Sakrī.* Subnaha.*

* Outworks of the Dhūli factory in the Muzaffarpur district.

CHAPTER X.

MEANS OF COMMUNICATION.

IN the early days of British administration North Bihār was far better provided with roads than the South Gangetic districts, owing to the European enterprise which was attracted to this part of the country. With an industry so valuable as the manufacture of indigo, it was only natural that attention should be given to the maintenance and improvement of means of communication; and the following extract from a report submitted by the Collector of Tirhut in 1824 shows that such roads as the interior of the district possessed were almost entirely due to the efforts of the planters. "The roads," he wrote, "are not under my control, but under that of the Magistrate, who usually repairs them with his prisoners as far as they can conveniently be sent. The zamindars do little or nothing in that way. The roads in the immediate vicinity of the different indigo factories are usually in good order, but they are kept so solely at the expense of the proprietors of the factories. The roads or bye-ways in the interior of the district are very bad and barely possible for hackeries. There are a number of streams and *jheels* in the district, wherefore bullocks are not much used: hackeries are, however, used near the indigo factories, and occasionally towards the northward, but the wear and tear of the same is great in consequence of the usually bad state of the roads. Owing to the number of streams, much of the produce of the district is conveyed by water. Where the roads are good, it is either owing to their being repaired by the Magistrate or by the planters." Little however appears to have been done to extend and increase the number of these roads, until a District Committee was established in 1870 for the administration of the funds which were set apart for the construction, maintenance and repair of roads, bridges, etc. During the famine of 1874 relief labour was largely employed in constructing new roads and repairing the existing ones; but in spite of this activity, the total length of roads made over to the Road Cess Committee of Darbhangā, on its formation in 1875, was only 648 miles. ROADS.

Since that date the construction of additional roads and the maintenance of old ones have received close attention; and in

1905-06 the length of metalled roads had increased to 52 miles, and of unmetalled roads to 1,953 miles, besides 766 miles of village roads. In other words, the length of roads is now more than thrice what it was 30 years ago, and there is more than half a mile of road to every square mile of superficial area, in spite of the fact that the resources of the District Board have been severely strained at times by the enormous damage caused by the floods which occasionally sweep down upon the district. Much has been done during the quinquennium ending in 1904-05 in the repair, raising and metalling of old roads, in the making of new roads, and in the construction of bridges. During this period the aggregate expenditure on original works has been 4½ lakhs or 61 per cent. more than in the preceding 5 years, and that on repairs 3⅔ lakhs. Special attention has also been paid to the extension of roads as feeders to the railway, and the length of the district roads has been increased by 46 miles.

Roads radiate from Darbhangā town and the subdivisional head-quarters to the most important places in the interior, and from Darbhangā town and the Sakrī, Jhanjhārpur and Nirmāli railway stations to the Nepāl frontier. The most important of these roads is the main road running eastwards from Muzaffarpur through Darbhangā town and Narahiā to Purnea. The central and south-eastern portion of the district is still, however, in want of good communications, and many of the roads are impassable during the rains. The tract is a low-lying one, subject to frequent floods, and high embankments are a necessity. These are expensive to maintain in repair, and are often breached in time of flood. To avoid this, an enormous amount of bridging would be necessary. Much, however, has been done in this direction in recent years; five pontoon bridges have been constructed at different points on the Gandak and the Bāghmati; and the road from Darbhangā to Jaynagar on the frontier, which crosses all the rivers in the west of the Madhubanī subdivision, has been bridged throughout at the cost of the Darbhangā Rāj. In Samāstipur, where the country is high, and comparatively little embanking or bridging is required, most of the roads are in good order and can be used at all seasons of the year.

Roadside arboriculture. There are few, if any, old roads on which portions of avenues of trees do not exist; but for some years previous to 1900 nothing was spent on tree-planting. It was represented that the trees damaged the roads, absorbing the moisture in the dry weather and preventing the roads from drying up during the rainy season. It was admitted, however, that this argument could not apply to the metalled roads, and roadside arboriculture has been resumed, the efforts of the District Board being directed mainly to maintaining avenues where they exist and to planting new trees along these roads. Since 1900 a 12 years programme has been in force, under which the Board devotes at least Rs. 1,000 a year on the average to this object. The great difficulty in the way of the scheme has been the supervision of the work at any distance from head-quarters; and the Board has therefore encouraged private enterprise as much as possible. The conditions observed are that the trees, when planted, become the property of the Board as long as they exist; and when they fall or are cut down, the timber vests in the man who planted them. It is claimed that the good results obtained have justified this system, the trees receiving more attention in their infancy, and the finances of the District Board being relieved.

The two kinds of carts most commonly used for transporting goods are the *chaghūs* and the *saggar*. Both are two-wheeled carts drawn by bullocks, but the former is a large cart with long bamboo poles projecting in front from either side, and the latter is a lighter and rougher kind of cart, usually employed for carrying country produce. The conveyances most generally used by passengers are the *ekkā*, *manjholi* and *champani*. The *ekkā* is a two-wheeled light trap, drawn by a single pony, which can be used over the most uneven ground; the body consists of a frame-work, covered with cloth, across which *newār* tape is woven. The *manjholi* and *champani* are both drawn by 2 bullocks. The former is similar in construction to an *ekkā*, but the yoke consists of a beam of wood at right angles to another long beam projecting from the body of the cart. The *champani* is a two-wheeled and sometimes four-wheeled light carriage similar in construction to an omnibus. It has however no benches, and the travellers squat or lie down as they please. It has a pole with a cross-bar, which rests on the backs of the bullocks which drag it. Carts.

The district is, on the whole, well provided with railway communications. Its south-west corner is traversed for 29 miles by the main line of the Bengal and North-Western Railway, which enters the district just below Dalsingh Sarai and runs through Samāstipur to Waini; and it also contains 25 miles of the chord line from Hājīpur to Bachwāra, which runs parallel to the Ganges embankment from east to west. From Samāstipur a line runs to Darbhangā and there branches off in two directions, the first north-west to Sītāmarhi through Kamtaul and Jogiāra, and the other due east to Khanwā Ghāt on the Kosi. The total length of these lines within the district is 146 miles; and another line running from Sakrī to Jaynagar on the Nepāl frontier has recently been RAILWAYS.

opened. This line, the earthwork of which was mostly completed as a relief work in the famine of 1897, has opened up a tract which was previously remote from the railway; it passes the important town of Madhubanī and taps a large grain supply from Nepāl. The construction of a direct line from Darbhangā to Muzaffarpur with a branch line to Sītāmarhi has also been sanctioned.

The District Board of Darbhangā have proposed the construction of a light railway from Samāstipur to Roserā, and the District Board of Monghyr desire to have a similar railway from Roserā to Khagariā. As, however, the Bengal and North-Western Railway Company has proposed a line from Darbhangā to Khagariā, which will serve much of the country to be traversed by the above two lines, and as it yet remains to be seen what further railway connections the proposal will involve, both the light railway projects are held in abeyance for the present. The Railway Board has now (1906) sanctioned the survey of the line from Darbhangā to Khagariā and also of a line from Darbhangā to Roserā and Khagariā, both surveys being made by the agency of the Bengal and North-Western Railway.

WATER COMMUNICATIONS. The Ganges is navigable for steamers throughout the year, and a daily service which plies up the river from Goalundo calls at Hardāspur in the extreme south-west corner of the Samāstipur subdivision. The Little Gandak river is navigable for boats of 1,000 maunds burden at all seasons, but its boat traffic has much decreased since the opening of the railway. Boats of 400 or 500 maunds can pass up the Bāghmati, except in a very dry season. The other rivers in the district are navigable in the rainy season only, but are not much used even then owing to their liability to floods. The principal ferries are those on the Little Gandak and Bāghmati rivers, the most important being at Māgardihi Ghāt (at Samāstipur) and Singiā Ghāt (at Roserā) on the Little Gandak, and at Kalyā Ghāt and Hayā Ghāt on the Bāghmati. A steam ferry also plies between Bāzidpur and Bārh in the Patna district on the main line of the East Indian Railway.

POSTAL COMMUNICATIONS. There are altogether 414 miles of postal communication and 62 post offices in the district, there being thus one post office for every 54 square miles. The number of postal articles delivered in 1905-06 was 2,243,000; the value of money orders issued was Rs. 14,57,000, and of those paid Rs. 19,14,000; and the total amount of Savings Bank deposits was Rs 1,61,000. There are also 13 telegraph offices, from which 17,500 messages were issued in the year; these offices are situated at Dalsingh Sarai, Darbhangā, Kamtaul, Laheriā Sarai, Madhubanī, Narahiā, Narhan, Pandaul, Pūsa, Rājnagar, Roserā, Samāstipur and Tājpur.

CHAPTER XI.

LAND REVENUE ADMINISTRATION.

GENERAL REVENUE HISTORY.

THE first historical assessment of land revenue in Tirhut is that made in 1582 by Todar Mal, the great finance minister of Akbar. The result of this assessment was that an area of 817,370 acres in *sarkārs* Hājīpur and Tirhut was settled at a revenue of Rs. 11,63,020, which gives an incidence of about Re. 1-7 per acre. Owing to the changes of jurisdiction which have taken place, it is unfortunately impossible to disentangle the figures for the present district of Darbhangā. During the period, close on 200 years, which elapsed between this assessment and the grant of the *Dīwāni* in 1765 to the East India Company, there are two assessments of which we possess statistical knowledge, one made by Shāh Jahān in 1685 and the other in 1750 during the viceroyalty of Alī Vardi Khān. The result of the first assessment was to raise the revenue to Rs. 17,98,576, *i.e.*, by 55 per cent., while it was lowered by Alī Vardi Khān to Rs. 16,48,142. No records can, however, be found of the areas dealt with, so that we are unable to estimate the incidence upon the actual area cultivated at the time they were made. It appears to have been difficult to ascertain what was the actual state of the revenues when we took over the Province in 1765, and various changes were made in the system of revenue administration up to the date of the decennial settlement in 1790, which was made permanent by Lord Cornwallis in 1793. At this settlement an area of 1,584,836 acres was assessed in Tirhut at Rs. 9,83,642, giving an incidence of 9 annas per acre. During subsequent resumption proceedings a further area of 1,066,001 acres was assessed to a revenue of Rs. 6,77,387, making a total of Rs. 16,61,029. The actually permanently settled revenue of Muzaffarpur and Darbhangā at the present time (1906) is Rs. 17,45,924.

EARLY ENGLISH ADMINISTRATION.

When the English assumed the government of the country in 1765, they continued the existing system of revenue administration, which the mercantile training of the Company's servants had not qualified them to manage. In 1769 European Supervisors were appointed to control the native officers, and, as it was realized that the collection of details regarding the internal resources of the country was a matter of the first importance, they wereo rdered to enquire into the economic conditions prevailing and the general system of administration. The result of these enquiries was to show that "the whole system resolved itself on the part of the public officers into habitual extortion and injustice, which produced on the cultivators the natural consequences, concealment and evasion, by which Government was defrauded of a considerable portion of its just demands." In 1770 a Revenue Council of Control was established at Patna; and next year the Court of Directors sent out their well-known orders "to stand forth as Dīwān and by the agency of the Company's servants to take upon themselves the entire care and management of the revenues." The direct control of the revenue administration was, accordingly, entrusted to European officers; but this assumption of direct management having proved a financial failure, the European agency was replaced by native *āmils* under the control of a Provincial Council at Patna. The quinquennial settlement effected in 1772 having also failed to give satisfaction, annual settlements were made from 1778-80 with farmers and zamīndārs; and these settlements being equally unsuccessful, Tirhut was for the first time placed under an European Collector in 1782.

The first Collector of Tirhut. The first Collector of Tirhut was Francois Grand*, whose career deserves at least a brief notice. A native of Lausanne, he was educated in his own country, but was sent in early life to England, where he obtained a cadetship in Bengal. He served in the forces of the East India Company until he obtained a captaincy, and in 1776, while on furlough in Europe, he was nominated to a writership. On his return to India, he married the Danish beauty, Mademoiselle Worlée, who afterwards as Madam Grand became so well known for her liaison with Sir Philip Francis and her subsequent marriage with Talleyrand. In 1779 Grand brought an action against Francis for misconduct with his wife, and was awarded damages amounting to Rs. 50,000 by the Judges Impey, Chambers and Hyde. He then obtained a divorce, and in 1782 was appointed Collector of Tirhut, where he engaged in, and practically founded, the indigo industry. He quickly began to make a fortune, but in 1787 Lord Cornwallis, hearing of his commercial enterprises, removed him from Tirhut. He was appointed Judge and Magistrate at Patna, warned to give up his indigo concerns, and ultimately removed from the

* This sketch of this adventurer is based on the article on Grand in Buckland's Dictionary of Indian Biography.

service. After this, Grand returned to Europe and made his way to France, where his former wife had married Talleyrand and risen to the position of Princesse de Benevento. Through the influence of Talleyrand, he obtained from the Batavian Republic in 1802 the post of Privy Councillor to the Government of the Cape of Good Hope; and later when the country passed to the British, he took service under them, with an easy adaptability which seems to have been natural to him. He was appointed Inspector of Woods and Lands, married again, and died at the Cape in 1821.

The foregoing sketch sufficiently indicates the character of the first Collector of Tirhut, a foreign adventurer with few scruples and but little sense of honour. The following report written by him in 1785 is of some interest as giving an insight into the state of the country and his method of exercising at one and the same time his administrative and commercial abilities: "In 1782 I was transferred by Mr. Hastings from Head Assistant to a commercial factory (in which the duties consisted of prizing cloths, seeing saltperte weighed or loaded, attending to the accounts, etc.) to the Government of two considerable provinces involving the settlement or collection of revenues and maintenance of justice: the provinces were Tirhut and Hajipur. I took possession of a country yielding a revenue of above seven lakhs of rupees, but which had suffered from the depredations committed by those who were compelled to abandon the charge to me, and had besides been in revolt owing to the intrigues of the Raja of Benares, Chet Singh, whose baneful influence had spread so far and would have spread further had he not been checked in time by Mr. Hastings' wise and spirited measures. I recovered a large balance due from the farmers to Government, quieted and appeased without bloodshed every disturbance, brought back the disobedient to a just sense of their errors, augmented the revenue, introduced the manufacturing of indigo after the European manner, encouraged the establishment of indigo works and plantations, erected three at my own expense, and thus possessed a fortune of £15,000 sterling, looked forward to a proportionate augmentation by continuing in my station and extending my manufactories, which with my houses, lands, furniture, tents, equipages, horses, boats, stood upon a valuation of £10,000 more."

This extract shows the high opinion Grand had of his own abilities, but the protest he made when removed from Tirhut displays even more clearly his inordinate self-conceit. In this he wrote:—"On the 26th August 1787 I was in full possession of my appointment, and my fortune was in that progressive state as described in 1785. I was in the enjoyment of every comfort, elegance and luxury of life. I was beloved and respected by those living with me, and I will say, because I challenge the contrary to be proved, almost vernerated by the natives of every description under my government, whose tears on hearing of my removal accompanied me even from the place of my residence to the bank of the Ganges, where the limits of the district ceased, a distance of 25 miles. On the 27th August 1787, by one stroke of His Lordship's pen, was Mr. Robert Bathurst nominated Collector of Tirhut and Hajipur, and thus every hope and fair-built prospect existing on the preceding day completely blasted. Thus the blow was struck, and from that day I fell, perhaps never more to rise. View the portrait and feel!"

Early settlements. The history of revenue administration from 1781 onwards is marked by yearly settlements with zamīndārs, and, if they were recusant, with farmers. The correspondence of the period is full of the difficulties the Collector had to meet both in settling and realizing the revenue. Himself ignorant of the capabilities of the soil, his subordinates were a hindrance rather than a help; and the enormous number of petty proprietors, for which the district was even then notorious, made a fair settlement a very difficult task, as they quarrelled and haggled over every detail. The only large proprietor, Rājā Madhu Singh of Darbhangā, contumaciously refused all terms, and did his best, by intimidating the farmers, to prevent them taking settlement of his estates or from collecting rents, if they did take settlement. Imperfect as was the assessment made under such circumstances, the difficulties of collecting it were aggravated by the lawless state of the country, which was infested by bands of robbers, who were generally in collusion with the native *āmils* and did not hesitate on occasion to molest even the European servants of the Company. To the trouble caused by depredating land-owners and contumacious Rājās was added that caused by unfavourable seasons, as scarcely a year passed without the record of some natural calamity. In one year it was drought, in the next it was flood, but in either case the result was the same, settlement holders unwilling to fulfil their engagements without extreme coercion.

Permanent Settlement. The history of Tirhut from this time forward until the declaration of the Permanent Settlement discloses persistent and more or less systematic attempts to acquire information regarding the capabilities of the soil, the relations of landlord and tenant, their status, and similar matters of rural economy; but the methods of enquiry were not sufficiently detailed to afford an adequate basis for a settlement in perpetuity. Accordingly, when the decennial settlement was declared permanent in 1793, no less than three-fifths of the area of Tirhut escaped assessment. The revenue actually fixed was, however, 15 per cent. more than in 1787; and the basis of assessment was more accurate than in Bengal, as the facilities for obtaining information were greater and the organization of village officials was still effective. The result was that, whereas in Bengal the Permanent Settlement served to confirm as zamīndārs and land-owners men who had been mere rent collectors under previous administrations, it served in Tirhut to rescue the real zamīndārs from the farmers of revenue who had been placed over them, and to restore proprietary rights to those who were about to lose them. On the other hand, incalculable injury was done to the ryots by the absence of any accompanying measures for their protection, and by the harsh provisions of Regulation VII of 1799, which extended to the landlords of Tirhut, as well as of the rest of the Province, the power to distrain the crops of their ryots, and in certain cases to arrest their persons for arrears of rent without any reference to the Courts. Whatever justification there may have been in Bengal for enhancing the power of the landlords in this way, there is no indication that such stringent measures were required in Tirhut, where the revenue fixed at the Permanent Settlement was collected with great success, and the number of estates in arrears was never so large as to endanger seriously the Government revenue. The country has, accordingly, suffered up to the present day from the failure of the earlier administrators to gauge adequately the extent of the oppression practised by the landlords of Tirhut upon their unfortunate tenantry. By slow degrees, the Legislature has been signifying by successive enactments its recognition of the rights of the cultivators; but it was not till the recent introduction of survey and settlement proceedings in Darbhangā that the acknowledgment of their rights was brought home to the people themselves.

RESUMPTION PROCEEDINGS.

The principal features in the subsequent revenue history of the district are the resumption of invalid revenue-free grants, the revenue survey of 1843-49, the enormous increase in the number of estates, and the settlement operations of 1892-99. It has already been stated that at the time of the decennial settlement the area of estates assessed to revenue was less than half the area of Tirhut. A large number of estates escaped assessment under the claim that they were revenue-free grants, of which there was an exceptional number in Tirhut. The free grant of land to Brāhmans for their maintenance, for the encouragement of learning, or for the worship of the gods has always been recognized by Hindus as a becoming act of piety. It is not surprising, therefore, that in Mithilā, whose chief claim to a place in history rests on its former influence as a centre of Hindu religion and learning, rent-free grants to the learned and priestly caste were exceptionally numerous. Mithilā, though swept by the successive inroads of invasion and conquest, still retained much of its ancient power as a stronghold of Hinduism; and this was a characteristic of that religion the least likely to succumb to outside attack. Again, it was customary in the Mughal era to remunerate civil and military subordinates by grants of land; and the remoteness of the district from the centre of the Mughal power rendered it easy for the subordinate officials of the Empire to carve out for themselves *jāgīrs* and *nankars*. To such an extent did this practice grow that at the final Muhammadan assessment of 1765, more than 5 lakhs, out of a total demand of less than 8 lakhs of rupees, was retained by *jāgīrdārs* and other revenue-free holders, and never reached Government at all, the chief offender being the officer in charge of the assessment, who secured for himself property yielding a revenue of nearly 1½ lakh of rupees. The effects of the resumption proceedings on the revenue roll of Tirhut were, consequently, almost as important as those of the Permanent Settlement; and as the resumption was carried out with much more care than the Permanent Settlement, it proved to be a task of even greater magnitude. The validity of the claims put forward to such revenue-free grants was carefully examined between 1830 and 1850; and systematic operations were undertaken for the resumption of those grants that proved to be invalid. Their extent can best be judged by the fact that altogether 1,066,000 acres were resumed in Tirhut, the revenue was increased by Rs. 6,77,387, and 3,018 new estates were assessed to revenue. In Darbhangā alone they resulted in an assessment of 928 square miles and in an addition of Rs. 3,60,596 to the revenue of the district.

PARTITION OF ESTATES.

Since that time, advantage has been taken of the law of partition to a most remarkable extent, and by the year 1895 the number of estates on the revenue-roll of Darbhangā and Muzaffarpur was a little under 32,000. In 1790 the estates on the revenue-roll of the whole of Tirhut numbered only 1,331, and the aggregate of estates therefore multiplied nearly thirty-fold in a little over a century. It is true that resumptions helped to swell their number before 1850; but, apart from this, the law of partition was responsible for the addition of over 27,500 estates to the revenue-roll of 1790. Darbhangā accounts for a very large proportion of the increase. In 1878, when the district revenue-roll was finally separated from that of Muzaffarpur, the number of permanently settled estates was 7,962; there were 10,813 in 1889-90, and 13,400 in 1899-1900; and the number has now risen to 14,183 (1905-06). The corresponding area contained only 532 estates in 1790, and their number has therefore increased twenty-five fold.

Growth of revenue. The revenue of Darbhangā fixed at the Permanent Settlement was Rs. 5,47,512, but as the large estates of Rājā Madhu Singh were at that time not settled with him, the revenue shown for them was that at which temporary settlements were made with farmers. Little more than a quarter of the total area came under the operations; and nearly two-thirds of the present cultivated area escaped assessment altogether. Five years later the revenue had fallen to a little over 3 lakhs owing to the failure of the farmers to keep their engagements; and it continued to fluctuate in consequence of the temporary leases of Madhu Singh's estates, which were made at varying assessments until 1808. As a result of the resumption proceedings, no less than 45 per cent. of the present permanently settled revenue was added to the roll; but since the creation of the district the fluctuations in the annual demand have been insignificant. It has varied only from Rs. 7,89,093 in 1879 to Rs. 7,90,393 at the present day (1906); and the difference between the present revenue and that of 1795 is, therefore, about $4\frac{4}{5}$ lakhs, of which $3\frac{3}{5}$ lakhs are accounted for by the resumption proceedings, the difference of $1\frac{1}{5}$ lakhs being due to the settlement subsequently made with Rājā Madhu Singh. The revenue now fixed is payable by 14,193 estates, of which 14,183 with a demand of Rs. 7,85,632 are permanently settled; and practically all the remainder is accounted for by 8 Government estates, with a demand of Rs. 4,331. Besides these, there are 1,040 revenue-free estates, forming a little more than one-twentieth of the district area.

Incidence of revenue. The incidence of revenue amounts to only 5 annas 10 pies per acre, and is less than in any other district of North Bihār except Champāran. Nearly one-third of the district was assessed during the resumption proceedings at an average rate of $9\frac{3}{4}$ annas per acre, and the only conclusion, therefore, is that the rest of the district was very fortunate in securing a light assessment at the time of the Permanent Settlement. The extreme lightness of the Government demand is the more apparent from the fact that the average rent rate of the district is Rs. 3-12-4 per acre, and 80 per cent. of the total area is cultivated. At the Permanent Settlement the Government revenue was fixed so as to allow the zamīndārs 10 per cent. of the assets as profits, but they now enjoy 88 per cent.; in other words, the Government revenue instead of forming nine-tenths of the assets is now only one-ninth of them.

SURVEYS AND SETTLEMENTS. The district was first surveyed in the course of the great revenue survey of Bengal. This survey, which was commenced in 1846 and completed in 1849, showed the area of Tirhut as 6,114 square miles with a land revenue of Rs. 14,62,548 or annas 5-11¾ per acre. It led to the discovery of land still remaining unassessed, but its chief value was that it almost completely put a stop to constant disputes about boundaries, which were a fruitful source of litigation and a serious administrative difficulty. In 1865–68 a survey of the Gangetic *diāras* was undertaken under Act IX of 1847, with a view to the readjustment of revenue rendered necessary by changes in the riparian area. This was the only professional survey which took place before the survey and settlement of 1896–1903. These operations were instituted with the object of protecting adequately the interests of all parties interested in the land. It was realized that for this purpose legislative measures alone were insufficient; that they should be supplemented by the preparation of a full record of rights founded on an accurate survey; and that such a record could alone remove the want of fixity of tenure and put a stop to the arbitrary enhancement of rent from which the ryots in North Bihār suffered.

LANDLORDS. The result of the operations has been to show that the area occupied by landlords and tenants is 2,871 square miles or 86 per cent. of the total area of the district. The area in the occupation of landlords is 11 per cent. of the whole; and in spite of the fact that a very large portion of it is owned by the Mahārājā of Darbhangā, more than 40,000 proprietary interests and 135,000 proprietors were recorded. In the Madhubanī subdivision there has been little subdivision of proprietary interests, owing to the predominating presence of the Darbhangā Rāj, but elsewhere subdivision has been very minute, and it may fairly be said that there are a host of petty proprietors. The average area in possession of a single proprietor in each village, accordingly, varies from 204 acres in the Phulparās thāna, where the Darbhangā Rāj predominates, to 8 acres in Samāstipur, where the proprietor is little more than an ordinary cultivator. There can be no doubt that the increasing subdivision of proprietary rights, which is going on over a large part of the district, has made it necessary for the petty proprietors to resort to direct cultivation to a much greater extent than was formerly the case. Now that little further extension of cultivation is possible, while subdivision of proprietary rights is still going on, the position of many a petty proprietor must be a serious one. Hitherto he has been able, as a rule, to meet the wants of his growing family and the inevitable subdivision of his property which this entails, partly by ousting his ryots from their lands and taking them into his own cultivation, and partly by raising the cultivators' rents. But the record-of-rights which has now been prepared will, to a large extent, prevent him from resorting to either of these expedients, and many petty proprietors must shortly reach a stage at which they will no longer be able to live on the proceeds of their landed property, and will be forced to take to other avocations. Nor from the ryot's point of view is this altogether to be regretted, for it is a proverb in the district that the small proprietor is the harshest landlord.

So far there is no indication that the land is passing to any extent into the hands of the money-lenders. Nearly half the district is in the hands of large proprietors, and here there is no transfer of proprietary rights; while transfers elsewhere are generally made to persons of the landlord class.

TENURE-HOLDERS.

The area in direct possession of tenure-holders is less than 3 per cent. of the occupied area. Though there are few degrees of subinfeudation, simple tenures play an important part in the agrarian economy of Darbhangā, over one-fifth of the district area being let out to tenure-holders of different kinds. Rent-paying tenures account for 12 per cent. of the total area, and nearly half are of a permanent nature; but these cover only about a seventh of the total area under rent-paying tenures. The rise in the value of land in modern times has rendered proprietors disinclined to alienate their rights permanently, except at excessive rates or on payment of a very high *salāmi;* and most of the permanent tenures are of ancient date.

Permanent tenures.

Temporary tenures.

Temporary tenures are most numerous in the Samāstipur subdivision, where considerably more than three-fifths of the area comprised by them is held by indigo factories. In the district, as a whole, it is satisfactory to find that more than half the area held under temporary tenures is in possession of indigo planters, who, as a rule, are considerate landlords and do not tamper with rents. This leaves less than 100,000 acres, or little more than one twenty-fifth of the total area of the district, in the hands of ordinary temporary tenure-holders or *thīkādārs*, many of whom are mere rent speculators, whose one object is to make as much as they can out of the village during the term of their lease. As, however, the area in their possession is so small, their powers for evil are comparatively limited. The majority of the tenures are simple farming leases, only 43 per cent. being given as security for loans.

A description of these mortgages has already been given in the preceding chapter.

Rent-free tenures.

Rent-free tenures are much more important in Darbhangā than in any other district in North Bihār, covering, as they do, about one-tenth of the total area. The most numerous are those granted for religious purposes, which account for more than three-quarters of the whole number, but cover little more than a quarter of the total area under rent-free tenures. Nearly three-fifths of the total area is found in the three thānas of Madhubanī, Phulparās and Khajaulī, where the large proportion is due to the maintenance grants given by the Mahārājās of Darbhangā to their relations in accordance with the custom of the family. The holders of such grants are allowed to hold the land rent-free subject to the payment of the Government revenue to the Mahārājā, but the grants revert to the Rāj on the death of the grantee without male heirs. Service tenures cover an insignificant area, most of the land held in this way belonging to petty village officials, such as *barāhils* and *goraits*, who cultivate themselves. *Mālikānā* tenures, consisting of lands held rent-free in perpetuity by former proprietors of the village, are also of little importance.

Under-tenures.

Under-tenures cover less than 2 per cent of the total area of the district. They are most numerous in the Khajaulī thāna, and are mainly sub-leases granted by the Madhubanī Bābus, who are themselves maintenance-holders under the Darbhangā Rāj. In the south, where great subdivision of proprietary rights prevails, both tenures and sub-tenures are comparatively unimportant, for the area held by each proprietor is so small that he cannot afford to sublet his rights in it.

TENANTS.

There are less than 5,000 ryots at fixed rates. and they hold only a little over 10,000 acres ; this low percentage being due to the large enhancements of rates which has taken place during the last century.

Occupancy N ryots.

Nine-tenths of the occupied holdings and over four-fifths of the occupied area belong to settled or occupancy ryots. In the Madhubanī subdivision the area and proportion of holdings held by occupancy ryots are still higher than this. Here the average size of their holdings is over two acres ; but in the rest of the district the larger area in the direct cultivation of proprietors and tenure-holders leaves less room for the ryots, and the average size of such holdings is only about 1½ acres. From special enquiries made during the last settlement it was found that a little over 6 per cent. of the occupancy holdings and only 2½ per cent. of the area held by occupancy ryots were affected by sales or mortgages with possession. Taking the district as a whole, it is satisfactory to find that four-fifths of the transferees are ryots, and that only a little over 11 per cent. are money-lenders. The latter class is most numerous in the head-quarters and Samāstipur subdivisions, where there are an exceptional number of petty proprietors, who are very often practically indistinguishable from ryots. In no part of the district are there any signs that land is passing to any considerable extent into the hands of profesional money-lenders to the detriment of the actual cultivators.

Non-occupancy ryots cultivate little more than a half per cent. of the occupied holdings. Except in Madhubanī, the average size of their holdings is greater than that of occupancy ryots, owing to the fact that certain indigo factories in Darbhangā and Samāstipur hold considerable areas as non-occupancy ryots. Non-occupancy ryots.

Under-ryots are an unimportant class, cultivating only 32,491 acres or little more than 2 per cent. of the ryoti area. Less than 7 per cent. of ryoti holdings are sublet, and most of these are only sublet in part. The average size of an under-ryot's holding is only a little over half an acre. There is thus no indication that ryoti interests are being acquired by non-agriculturists and sublet to the actual cultivators. The largest proportion of under-ryots is found in the Samāstipur subdivision, where indigo factories often hold land under this status on the *kurtauli* leases mentioned in the last chapter. Elsewhere, the under-ryot usually takes up for a short time some portion of a holding, which the original ryot is unable to cultivate entirely himself. Under-ryots.

The only peculiar tenancy in Darbhangā is that known as *jāidādi* (from *jāidād*, a crop), which is found in the Baherā thāna. In the *chaurs* or marshes which are characteristic of this part of the district, a crop can only be grown in an exceptionally dry year; and the cultivators of such lands pay rent only in the year in which a crop is grown, and only for such areas as produce a crop. These areas are measured at the time of harvest and pay either a cash rent at a rate previously agreed upon or a produce rent. For areas in which the crop, though sown, has been destroyed by floods, no rent is charged. These tenures resemble the *utbandi* tenures found elsewhere in Bengal, and at the last settlement the ryot was held not to have occupancy rights in such lands, unless he had held them continuously for 12 years. *Jāidādi* tenancy.

CHAPTER XII.

GENERAL ADMINISTRATION.

ADMINISTRATIVE CHARGES AND STAFF. THE administration of the district is in charge of the Collector under the Commissioner of the Patna Division, and for general administrative purposes it is divided into three subdivisions with head-quarters at Darbhangā (Laheriā Sarai), Madhubanī and Samāstipur. The head-quarters subdivision is under the direct supervision of the Collector, while each of the other two subdivisions is in charge of a Subdivisional Officer exercising the powers of a Deputy Collector in revenue matters. At Darbhangā the Collector is assisted by a staff of Deputy Collectors consisting generally of five officers. In addition to this staff, there are two officers engaged on special work, viz., a Special Deputy Collector in charge of excise and a Sub-Deputy Collector employed on partition work. A Sub-Deputy Collector employed in the general line is also usually posted there, and occasionally a Joint Magistrate and an Assistant Collector. Each of the Subdivisional Officers of Madhubanī and Samāstipur is assisted by a Sub-Deputy Collector.

REVENUE. The revenue of the district under the main heads was Rs. 14,30,000 in 1880-81 (when the income-tax had not been imposed), Rs. 18,18,000 in 1890-91, and Rs. 19,60,000 in 1901. In 1905-06 it amounted to Rs. 21,77,000, of which Rs. 7,90,000 were derived from land revenue, the other main heads of income being cesses (Rs. 5,47,000), excise (Rs. 3,94,000), stamps (Rs. 3,85,000) and income-tax (Rs. 61,000).

Cesses. The road and public works cesses are, as usual, levied at the maximum rate of one anna in the rupee. The current demand in 1905-06 was Rs. 4,95,440, the greater part of which (Rs. 4,61,880) was payable by 17,958 revenue-paying estates, while practically all the remainder was due from 1,040 revenue-free estates. The number of tenures assessed to cesses was 17,359, and there are thus nearly as many tenures as estates in the district; while the number of recorded shareholders of estates and tenures was 84,309 and 23,903 respectively.

The excise revenue is, as usual, derived from imported liquors, country spirits, *tāri*, opium, and the duty and license fees on hemp drugs. A statement of the various exciseable articles and of the sums realized from them in the decade 1893-1902 is given in the Statistical Appendix, from which it will be apparent that the income from this source rose from 2 lakhs in 1892-93 to 3 lakhs in 1900-01. Since that year there has been a steady growth in the receipts, and in 1904-05 they amounted to Rs. 3,72,000, the increase during the quinquennium being over 23 per cent. The increase was considerably above the average for the Patna Division, and the annual incidence of the excise revenue rose from anna 1-5 to anna 1-9 per head of the population; it was still, however, lower than in any other Bihār district. In 1905-06 the total receipts were even greater, amounting in all to nearly Rs. 3,94,000.

Drinking in Bengal is largely indulged in by Hindī-speaking races, aborigines and mixed tribes, and consumption also varies inversely with the proportion of Muhammadans in the district. Darbhangā is a Hindī-speaking district, but the number of Muhammadans and men of aboriginal descent is small. The result is that the people, as a whole, are not hard drinkers, and the amount spent on liquor is smaller in proportion to the population than anywhere else in Bihār. The greater portion of the excise revenue is derived from the sale of country spirit, which is prepared by distillation from the flower of the *mahuā* tree (*Bassia latifolia*) and molasses, and of the fermented palm juice called *tāri*. These receipts amounted in 1905-06 to Rs. 2,61,000, a figure representing an expenditure of only Rs. 896 for every 10,000 of the population as compared with the Divisional average of Rs. 1,910. The manufacture and sale of country spirit are carried on under what is known as the dual system, *i.e.*, there is a central distillery at Darbhangā and outstills for the supply of the rest of the district. The average consumption of outstill liquor is 14 and of distillery liquor 37 proof gallons per thousand of the population, the incidence of taxation per head of the population being 9 pies and 2 annas, respectively. There are 17 shops for the sale of distillery liquor and 84 selling outstill liquor, *i.e.*, one retail shop for the sale of country spirit to every 28,837 persons; and besides these, there are 1,080 shops licensed to sell *tāri* or one shop to every 2,696 persons.

The receipts from hemp drugs account for nearly the whole of the remaining excise revenue, and are greater than in any other district in the Division, except Shāhābād. In 1905-06 they amounted to Rs. 1,20,000, all of which was obtained from the duty and license fees on *gānja*, *i.e.*, the dried flowering tops of the cultivated female hemp plant (*Canabis sativa*) and the resinous exudation on them. Only Rs. 10,700 are derived from the consumption of opium, and Rs. 1,400 from the license fees on imported liquors. The latter have found no favour with the mass of the population, both because they are unable to afford them, and because they prefer the country spirit and *tāri* they have drunk for generations past.

Stamps. The revenue from stamps ranks next in importance as a source of income to that derived from excise. During the ten years ending in 1905-06 it increased enormously, rising from Rs. 2,16,000 to Rs. 3,85,000. The increase is mainly due to the growing demand for judicial stamps, which brought in Rs. 3,07,000, as compared with Rs. 1,45,000 in 1895-96; the receipts from this source have thus been doubled in the last ten years. The sale of court-fee stamps, which in 1905-06 realized Rs. 2,80,000, is by far the most important item in the receipts from judicial stamps. Among non-judicial stamps, impressed stamps account for Rs. 73,600 or nearly the whole of the receipts under this head.

The sale of court-fee stamps grew by nearly 80 per cent. in the quinquennium ending in 1904-05, an increase greater than in any other Bihār district. The increase in their sale has been caused by the general growth of litigation which has occurred since the completion of the survey and settlement operations. The sale of impressed stamps, on the other hand, did not increase, and the total receipts during the quinquennium were slightly less than in the preceding 5 years. This small decrease is possibly due to the fact that land was not so much in demand as in the previous quinquennium, in which famine prevailed, driving many of the poorer classes to the money-lender.

Income-tax. From the Statistical Appendix it will be observed that in 1901-02 the income-tax yielded altogether Rs. 71,370 paid by 2,583 assessees, of whom 1,739 paying Rs. 19,690 had incomes of Rs. 500 to Rs. 1,000. At that time the minimum income assessable was Rs. 500, but this was raised in 1903 to Rs. 1,000 per annum, and the number of assessees consequently fell in 1903-04 to 993, the net collections being Rs. 60,650. In 1905-06 the amount collected was Rs. 61,000 paid by 988 assessees. Of these, 161, paying Rs. 20,000, are inhabitants of Darbhangā town, where the incidence of the tax is under a third of an anna per head of the population. The realizations are chiefly on account of grain and money-lending, the renting of houses and trade.

Registration. There are 9 offices for the registration of assurances under Act III of 1877. At Muzaffarpur the Special Sub-Registrar deals, as usual, with the documents presented there, and assists the District Magistrate, who is *ex-officio* Registrar, in supervising the proceedings of the Rural Sub-Registrars who are in charge of the other registration offices. The marginal statement shows the number of documents registered and the receipts and expenditure at each office in 1905. The average number of documents registered annually during the quinquennium ending in 1904 was 27,262 as against 27,938 in the preceding 5 years, the decrease amounting to 2·4 per cent. The cause of this decrease is partly that famine and distress in the previous quinquennium led to the registration of an exceptionally large number of documents, and partly that in the course of the settlement operations the zamīndārs in many cases took *kabuliyats* from their ryots to secure proof of the rents payable by them.

NAME.	Documents registered.	Receipts.	Expenditure.
		Rs.	Rs.
Darbhangā	5,440	8,882	3,860
Baherā	2,480	2,600	1,440
Benīpatti	2,887	2,347	1,378
Dalsingh Sarai	2,923	3,822	1,349
Madhubanī	4,585	4,569	1,808
Narahiā	4,390	3,664	2,010
Roserā	2,315	2,766	1,130
Samāstipur	4,018	4,568	1,851
Wārisnagar	1,929	1,855	1,002
Total ...	30,967	35,073	15,808

Administration of Justice. Until 1906 Darbhangā was included in the jurisdiction of the District and Sessions Judge of Muzaffarpur; but it has now been constituted a civil district and sessions division under a separate District and Sessions Judge, with head-quarters at Darbhangā.

Civil Justice. The judicial staff entertained for the administration of civil justice consists of the District Judge and 8 Munsifs, of whom three hold their courts at Darbhangā, three at Samāstipur and two at Madhubanī. Statistics of the civil work will be found in the Statistical Appendix, from which it will be seen that the number of cases tried by the Civil Courts rose from 4,559 in 1893 to 18,605 in 1902. This increase is due to the greater number of suits for arrears of rent, which amounted to 15,243 in 1902, as against an average of between 3,000 and 4,000 per annum prior to the year 1900. A special enquiry was made on the subject by the Special Judge appointed under the Tenancy Act; and the conclusions arrived at by him were that, although there had been some dissatisfaction on the part of the landlords with the rates of rent fixed by the Settlement Department, the increase was due chiefly to other facts, viz., that ryots withheld rent while the operations were in progress, in order to ascertain what rates would be settled, and that the landlords now have a record on which they can safely rely, in order to prove the relationship of landlords and tenants, whereas previously they were deterred from litigation by the fear that the papers filed by them would not be accepted as trustworthy by the Courts. The number of such suits has continued to increase, and in 1904, out of 21,460 civil suits disposed of by the Munsifs, no less than 18,000 were rent-suits.

Criminal Justice.

Criminal justice is administered by the District and Sessions Judge, the District Magistrate and the various Deputy and Sub-Deputy Magistrates at the head-quarters and subdivisional stations. The sanctioned staff at Darbhangā consists, in addition to the District Magistrate, of four Deputy Magistrates of the first class and one Deputy Magistrate of the second or third class. Besides these officers, an Assistant Magistrate and one or two Sub-Deputy Magistrates exercising second or third class powers are generally posted there. The Subdivisional Officers at Madhubanī and Samāstipur are almost invariably officers vested with first class powers, and they are usually assisted by Sub-Deputy Magistrates of the second class. There are also benches of Honorary Magistrates at Darbhangā (8 members), Madhubanī (9 members) Ruserā (11 members) and Samāstipur (3 members), all of which exercise second class powers. In all, there are 31 Honorary Magistrates, of whom nine are authorized to sit singly.

Statistics showing the work of the Criminal Courts will be found in the Statistical Appendix; and it will be sufficient to observe that the only district in North Bihār which showed an increase in criminal cases during the 5 years ending in 1904-05 was Darbhangā, in which the greater part of the survey operations were conducted during this period, and in which the leading zamīndārs at first showed a tendency to repudiate the decisions of the Settlement Officers.

Crime.

An account has been given in Chapter II of the lawless state of the country when it first passed under British rule. It was overrun by predatory hordes of banditti, and was infested by bands of robbers. This state of affairs has long since ceased, and dacoities are few in number, while robberies are very rare, those which do occur being generally of a technical kind; in the 5 years ending in 1904 only 2 dacoities and 10 robberies were committed. The people are peaceful and law-abiding as a rule, and serious offences and crimes of violence are the exception. Riots, connected with land disputes or with disputes arising out of cattle trespass or questions of irrigation, occasionally occur; but their number has fallen off remarkably since the completion of the survey and settlement operations. In the 5 years 1895–99 Darbhangā shared with Shāhābād the unenviable position of having more riots committed within its borders than any other district of the Patna Division, but in the next quinquennium the number of riots was less than in any of those districts, the annual average number of cases falling from 68 to 27. As in other parts of Bihār, thefts and burglaries are the commonest forms of crime. Burglary is especially common, and in the quinquennium ending in 1904 the number of burglaries in Darbhangā was greater than in any other district of the Division, except Patna and Gayā. The soft mud walls of the houses, the weary sleep of the inmates, the negligence, and often the acquiescence, of the village *chaukīdār* combine with the adroitness of the burglar to render his task an easy one and his arrest a comparatively rare occurrence. Further, the property stolen generally consists of brass utensils, trumpery ornaments, clothing, grain or cash; and when the same pattern prevails over the whole of the district, the identification of the property is as difficult as the concealment of it is easy.

POLICE. For police purposes the district is divided into 10 thānas or police circles, viz., Darbhangā, Baherā and Руserā, in the head-quarters subdivision; Madhubanī, Benīpatī, Khajaulī and Phulparās, in the Madhubanī subdivision; and Samāstipur, Dalsingh Sarai and Wārisnagar, in the Samāstipur subdivision. Besides these, there are 10 independent out-posts and 2 beat-houses, and there are thus 22 centres for the investigation of crime. The machinery employed for the protection of person and property consists of the regular or district police, including the town police, and of the rural force or village watch. The former consisted in 1905 of the District Superintendent of Police, 5 Inspectors, 37 Sub-Inspectors, 1 Sergeant, 32 head-constables and 416 constables; and the latter of 289 *dafadārs* and 4,530 *chaukīdārs*. The cost of the regular force was Rs. 1,03,000, and there was one policeman to every 6·7 square miles and to every 5,919 persons, as compared with the average of $4\frac{4}{5}$ square miles and 3,194 persons for the whole of the Patna Division.

JAILS. There is a district jail at Darbhangā and a subsidiary jail at each of the outlying subdivisional head-quarters of Madhubanī and Samāstipur. For a short time the subsidiary jail at Samāstipur was affiliated to the Muzaffarpur District Jail, but this arrangement did not work well and was given up some years ago. The jail at Darbhangā has accommodation for 355 prisoners, distributed as follows:—barracks without separate sleeping accommodation are provided for 244 male convicts, 15 female convicts, 16 juvenile convicts, 22 undertrial prisoners, and 7 civil prisoners; the hospital holds 33 persons; and there are separate cells for 2 Europeans and 16 male convicts. The subsidiary jail at Madhubanī has accommodation for 9 male and 3 female prisoners of all classes, and that at Samāstipur for 23 prisoners of all classes. The total daily average number of prisoners confined in the district jail in 1905 was 221, of whom 7 only were females; in the Madhubani sub-jail 12; and in the Samāstipur sub-jail 7.

CHAPTER XIII

LOCAL SELF-GOVERNMENT.

OUTSIDE municipal areas, the administration of local affairs is vested in the District Board and in the Local Boards, which have been constituted for the subdivisions of Madhubanī and Samāstipur. The District Board is responsible for the maintenance of roads, bridges and road-side rest-houses, and has the general superintendence of primary and middle class schools. It is also entrusted with the management of public ferries, the control over dispensaries, the provision of a proper water-supply, and village sanitation. To the Local Boards, which work in subordination to it, have been delegated the management of pounds and the administration of the sums allotted for the construction and repair of village roads.

THE DISTRICT BOARD.

The District Board was created in 1887 and consists of 25 members, of whom 12 are elected by the Local Boards, 8 are nominated and 5 are *ex-officio* members, viz., the Civil Surgeon, the senior Assistant Collector, the Road Cess Deputy Collector, the Deputy Inspector of Schools, and the District Magistrate, who is *ex-officio* Chairman. The Statistical Appendix shows, for the ten years 1892-93 to 1901-02, the principal sources from which it derives its income, and the objects on which it is spent; and it will suffice here to say that its average annual income during the decade was Rs. 3,57,000, of which Rs. 2,05,000 were derived from Provincial rates, while the average expenditure was Rs. 3,47,000, of which Rs. 1,85,000 were spent on civil works, Rs. 28,000 on education, and Rs. 5,000 on medical relief. During the quinquennium 1895-96 to 1899-1900 the average annual income was Rs. 3,47,000, and the expenditure Rs. 3,09,000, and in the next five years they were Rs. 3,60,000 and Rs. 3,10,000, respectively. It is the richest District Board in the Patna Division, and its income in the last quinquennium was 2 lakhs more than that of any other District Board in Bihār.

In 1904-05 the Board had an opening balance of Rs. 1,43,700, and its income from all other sources was Rs. 4,29,000. The expenditure in the same year was Rs. 3,49,000. Here, as elsewhere, the Provincial rates form the principal source of income, bringing in over 2½ lakhs; but the incidence of taxation is very light, amounting to only 1 anna 5 pies per head of the population. By far the largest portion of the income of the District Board is spent on civil works, *i.e.*, the extension and maintenance of communications, the up-keep of staging bungalows, the construction of buildings, and the provision of a proper water-supply. Altogether 2 lakhs were expended on these works in 1905-06, of which over three-fourths was spent on the construction, improvement and repairs of roads. The immediate administration of the roads is vested in the District Engineer, and that officer is also responsible for the management and repair of the dāk bungalow at Laheriā Sarai, 8 inspection houses, and other buildings belonging to the District Board. The Board also controls 113 ferries and 83 pounds, which are generally leased out to the highest bidder. The receipts from the latter source decreased from Rs. 6,712 in 1899-1900 to Rs. 6,158 in 1904-05; but on the other hand the revenue from ferries rose from Rs. 29,027 to Rs. 58,283. This large increase is due to better settlements and the acquisition of new ferries; but with an expanding railway system and with the construction of new bridges it can scarcely be sustained.

After civil works, education constitutes the heaviest charge on the resources of the District Board, entailing an expenditure of Rs 50,000, or nearly one-seventh of its total expenditure, in 1905-06. There is an inspecting staff of 7 Sub-Inspectors of Schools and 14 Inspecting Pandits, subordinate to the Deputy Inspector of Schools, and it maintains 1 Middle school and aids four others, besides aiding 1 Industrial school, 20 Upper Primary and 1,087 Lower Primary schools. It also awards scholarships tenable at the Bihār School of Engineering, and pays the stipend of a student at the Bengal Veterinary College at Belgāchia.

For the relief of sickness, it maintains 8 dispensaries, and aids four others; and when cholera breaks out in the interior, it despatches native doctors with medicines to the affected villages. The sanitary work done by the District Board is of a somewhat varied character. It includes preventive measures against plague, cholera, and other epidemics, sanitary arrangements at fairs, the construction, repair, and improvement of wells, and experiments in village sanitation, such as the clearance of jungle, the excavation of drains, and the filling up of hollows containing stagnant water. During 1905-06 it spent 4·2 per cent. of its ordinary income on medical relief and sanitation. It also contributes towards the maintenance of the veterinary dispensary which is kept up by the municipality at Laheriā Sarai, and maintains a staff of Veterinary Assistants for the benefit of the rural population, these officers being deputed, when necessary, to cope with outbreaks of epidemic disease in the interior.

It is reported that the District Board administers the duties entrusted to it satisfactorily and without friction. The members evince great interest in their work, and are always ready with valuable suggestions and advice. In particular, they are especially useful in making known the wants of their part of the district, in supervising repairs to the roads, and in controlling the working of the dispensaries and pounds.

LOCAL BOARDS. Subject to the control of the District Board, the Local Boards of Madhubanī and Samāstipur have jurisdiction over each of the subdivisions of the same name. There was formerly a Sadar Local Board for the head-quarters subdivision, but it was abolished some years ago, and the functions it discharged were transferred to the District Board. The Local Board of Madhubanī consists of 11 members, and the Local Board of Samāstipur of 8 members; the system of election which obtains in many Bengal districts has not been introduced, and the members of each Local Board are nominated, with the exception of one *ex-officio* member, the Subdivisional Officer, who is its Chairman. The duties of the Local Boards are confined to the management of village roads and pounds and village sanitation; and their chief duty is to maintain in an efficient state of repair the roads made over to them. The average expenditure in Madhubanī is Rs. 6,700 and in Samāstipur Rs 11,500 per annum.

MUNICIPALITIES. There are four municipalities in the district, viz., Darbhangā, Madhubanī, Ruserā, and Samāstipur. The total number of rate-payers is 15,762, out of a total urban population of 98,490, the ratio being 15·8 per cent. as compared with the Divisional average of 18·1 per cent. In all the municipalities taxation assumes the form of a tax on persons; the incidence of taxation varying from 6 annas 6 pies in Ruserā to 10 annas 7 pies in Darbhangā. It is, on the whole, very light, being 10 annas per head of the population as against the Divisional average of 12 annas 11 pies. Statistics of the receipts and expenditure in each municipality during the years 1892-93 to 1901-02 will be found in the Statistical Appendix.

The maintenance of roads, the improvement of drainage, and the provision of an adequate system of conservancy are the chief demands on the municipal income. Steps are being taken to supply these requirements, but of late years the resources of the municipalities have been strained by plague, which has hindered the execution of these necessary reforms. Still, much has been done recently to supply the principal civic requirements, and at the same time to maintain financial equilibrium. In Darbhangā the former municipal out-door dispensary has been converted into an important in-door hospital, a lady doctor has been entertained, and a veterinary dispensary has been established. A complete drainage scheme, costing nearly two lakhs, has been prepared, and is gradually being carried out; and the sanitation of the town has been considerably improved. In Madhubanī the question of drainage has been taken up, and an important channel has been constructed, which is designed to carry off water, to prevent the accumulation of stagnant pools, and to remove the danger of floods to which the town is exposed. Steps have also been taken to organize a fire-brigade among the municipal employés, and thus avoid the risks of fire. The Rusera Municipality has had the town surveyed and a complete record of municipal government prepared; and in Samāstipur, where the chief defect at present is the want of a complete system of drainage, levels of the town are being taken with the object of preparing an effective scheme.

Dar-bhangā. The Darbhangā Municipality, which was established in 1864, is administered by a Municipal Board consisting of 22 Commissioners, of whom 14 are elected, 4 are nominated, and 4 are *ex-officio* members. The area within municipal limits is 7 square miles divided into 7 wards, and the number of rate-payers is 9,785, or 14·7 per cent. of the population. The average income during the five years 1895-96 to 1899-1900 was Rs. 51,960, and the expenditure Rs. 37,870; in the next quinquennium they were Rs. 50,500 and Rs. 41,550, respectively. In 1905-06 the income of the municipality was Rs. 68,000, the chief source of income being a tax on persons, assessed, according to circumstances and property, at 1 per cent. of the income of the assessees. This tax brought in Rs. 22,700; and the total incidence of taxation was 10 annas 7 pies per head of the population. The expenditure amounted to Rs. 55,700, of which 28·5 per cent. was spent on conservancy, 24·8 per cent. on public works, and 17·1 per cent. on medical relief.

Madhu-bani. Madhubanī was constituted a municipality in 1869, and has a Municipal Board consisting of 16 members, of whom 10 are elected, 5 are nominated, and one is an *ex-officio* member. The area within municipal limits is 4 square miles divided into 5 wards, and the number of rate-payers is 2,725, or 21·1 per cent. of the total population. The average income during the five years

1895-96 to 1899-1900 was Rs. 17,840, and the expenditure Rs. 13,460; and in the next quinquennium they were Rs. 17,110 and Rs. 12,600 respectively. In 1905-06 the income of the municipality was Rs. 21,380, the principal receipts being Rs. 6,350 realized from a personal tax, assessed, according to circumstances and property, at ¾ per cent. of the income of the assessees; the incidence of taxation was 9 annas 7 pies per head of the population. The expenditure amounted to Rs. 13,800, the principal items being public works, which accounted for 29·1 per cent. of the expenditure, medical relief (28·6 per cent.), and conservancy (14·2 per cent.).

Ruserā. Ruserā was constituted a municipality in 1869, and has a Municipal Board consisting of 14 Commissioners, all of whom are nominated. The area within municipal limits is 2½ square miles, and the number of rate-payers is 1,806 or 17·6 per cent. of the population. The average income during the five years 1895-96 to 1899-1900 was Rs. 6,015, and the expenditure Rs. 5,210; and in the next quinquennium they were Rs. 6,220 and Rs. 5,520 respectively. In 1905-06 the income of the municipality was Rs. 6,430, of which Rs. 3,030 were realized from a tax levied according to the circumstances and property of the assessees; the incidence of taxation was 6 annas 6 pies per head of the population. The expenditure amounted to Rs. 5,500, of which 23·7 per cent. was spent on medical relief, 21·3 per cent. on general establishment, and 19 per cent. on conservancy.

Samāstipur. Samāstipur was constituted a municipality in 1897, and has a Municipal Board consisting of 9 Commissioners, of whom two are *ex-officio* members of the Board and the rest are nominated. The area within municipal limits is 1¾ square mile, and the number of rate-payers is 1,446 or 15·8 per cent. of the population. The average income during the 3 years 1897-98 to 1899-1900 was Rs. 4,190, and the expenditure Rs. 3,440; and in the next five years they were Rs. 10,160 and Rs. 9,450 respectively. In 1905-06 the income of the municipality was Rs. 11,150, of which Rs. 4,270 was derived from a tax on persons, assessed, according to their circumstances and property, at 12½ annas per cent. on the income of the assessees; the incidence of taxation was 10 annas 1 pie per head of the population. The expenditure was Rs. 9,440, of which 20·5 per cent. was spent on medical relief, 17·2 per cent. on education, and 17·5 per cent. on conservancy.

CHAPTER XIV.

EDUCATION*.

PROGRESS OF EDUCATION

THE advance of education among the people has been very rapid since the formation of the district in 1875. In that year there were 201 schools of all classes attended by 4,668 pupils; in 1885 the number of schools had risen to 2,532 and the attendance to 38,958; but there was a decline in the number both of schools and of pupils in the next decade, the former falling in 1895 to 1,456 and the latter to 29,578. They rose again however by the year 1905 to 1,734 and 45,457 respectively: in other words, the number of children receiving instruction has grown by over 50 per cent. during the last ten years, while the schools have increased by nearly 20 per cent., a rate of progress which is sufficiently satisfactory. The percentage of children under instruction to those of school-going age is still, however, only 10·4†, and there is approximately one school to every 2 square miles and to every 2 villages. The supervision of these schools is entrusted to an inspecting staff consisting of a Deputy Inspector of Schools, 5 Sub-Inspectors of Schools, 2 Assistant Sub-Inspectors and 14 Guru Inspectors.

SECONDARY EDUCATION.

High Schools.

There is no college in the district. There was only one High school during the first two decades, but the number rose to 5 in 1895 and again to 6 in 1905; and there was a corresponding increase in the number of pupils, which was only 352 at the end of the second decade, but rose to 1,005 in 1895 and to 1,180 in 1905. Of these six schools, four are in Darbhangā town, viz., the Northbrook School, the Darbhangā Rāj School, the Saraswati Academy and the Bengali High English School; and two are situated in the subdivisional head-quarters, Samāstipur and Madhubanī. The Northbrook School has 249 pupils, and is maintained at a cost of Rs. 6,400 a year; the Rāj School has 264 pupils and costs Rs. 5,860 per annum; the Saraswatī Academy has 195 pupils, and

* I am indebted to Mr. W. Billing, Inspector of Schools, Patna Division, for an account of education in Darbhanga from which this chapter has been prepared.

† The proportion of boys under instruction to the total number of boys of school-going age is 20 per cent. and the proportion in the case of girls is a little over 1 per cent.

the annual expenditure amounts to Rs. 3,450; and the Bengali School, which is intended mainly for Bengali pupils and only teaches up to the 4th class of a High English school, as a branch of the Rāj High English School, had 38 pupils on the rolls on the 31st March 1905, the expenditure for the year being Rs. 700. The Watson High English School at Madhubanī, which has 246 pupils, receives a grant-in-aid from Government, and is maintained at an annual cost of Rs. 4,860; while the Samāstipur High English School, which is attended by 188 pupils, is supported by fees only, which amounted in 1905 to Rs. 3,177. The average cost of educating each pupil is Rs. 25·7 per annum in the Zilla school and Rs. 22·2, Rs. 17·6 and Rs. 18·6 respectively in the Rāj School, the Saraswatī Academy and Bengali High English School, which are all private institutions. In Madhubanī, where the High school is aided by Government, the cost is Rs. 19·7, and in the Samāstipur High English School, which is unaided, it is Rs. 16·9 per annum.

Middle English Schools. The second class of secondary schools consists of the Middle English schools, *i.e.*, schools teaching up to the Middle Scholarship examination, in which English forms part of the recognized course of studies. The number of these schools increased from 6 in 1880 to 7 in 1885, declined to 4 in 1895, but again rose to 8 in 1905. The attendance similarly declined from 666 in 1885 to 323 in 1895, but rose again to 445 in 1905. All the schools of this class are in the interior of the district. One at Samāstipur, though aided by the District Board, is maintained by the municipality; three at Dighrā, Rahikā and Jogiārā are aided by the District Board; and the remaining three at Narhan, Dalsingh Sarai and Tājpur are maintained entirely by the Narhan estate.

Middle Vernacular Schools. The number of Middle Vernacular schools, *i.e.*, schools which read up to the Middle Scholarship, but in which the vernacular is the only recognized medium of instruction and forms the only course of studies, was 22 in 1880, and has declined steadily ever since, falling to 12 in 1885, to 9 in 1895, and finally to 6 in 1905. All these schools are situated in the interior of the district, and are maintained entirely by the Darbhangā Rāj. Not only the number of schools but also the attendance at them has been steadily declining, falling to 699 in 1885, to 438 in 1895, and again to 336 in 1905. The constant decrease in the number of schools, as well as in that of the pupils attending them, accompanied, as it has been, by a corresponding rise in the number and attendance of the High and Middle English schools, conclusively proves a growing preference for English education among the people,—a preference which is presumably due both to the intrinsic worth of the latter and to its higher market value, which make parents consider it an absolute necessity for their children. There are 8 Middle Scholarships allotted to the district, of which one is reserved for the Muhammadans.

PRIMARY EDUCATION.

In 1880 there were 507 Primary schools with 10,533 pupils in attendance; and the numbers rose to 1,634 and 28,406 respectively, in 1885, on account of the introduction of a system of payment by results. Consequent upon the strict enforcement, between the years 1885 and 1895, of rules prohibiting the recognition of and grant of aid to *pāthshālas* having 5 boys only or only such boys as were related to the teachers, the number of schools fell to 811 in 1904 and that of pupils to 21,738, but they rose to 1,190 and 36,773 respectively in 1905. The doubling of the number of pupils attending Primary schools is partly due to the grants from public sources having increased from Rs. 14,433 in 1895 to Rs. 38,614 in 1905, and partly to the introduction of the new scheme of vernacular education, which has made the *pāthshālas* more attractive. Of 36,773 pupils in Primary schools on the 31st March 1905, 31,329 were Hindus and 5,444 were Muhammadans. The cost of educating each pupil was Re. 1-0-10 from public funds and Re. 1-14 from all other sources. There are 5 Upper Primary and 34 Lower Primary Scholarships for boys.

SPECIAL SCHOOLS

There was no special school in the district prior to 1882, when the Madhubanī Sanskrit school with 59 pupils was for the first time entered in the departmental returns. The number of such schools has now risen to 49 and the attendance at them to 1,082 pupils. Of these, three are Guru Training schools, one at the head-quarters of each of the three subdivisions, which have been established for the training of the *gurus* of primary schools; one at Patahi is a female *guru* training school where females are trained for the purpose of teaching in the girls' schools of the district; one is an industrial school, opened for the training of carpenters, blacksmiths and painters on modern lines; and 44 are Sanskrit *tols* with 981 pupils. The most important of the latter institutions are the Narāyanī Pāthshāla at Akhtiārpur in the Samāstipur subdivision, at which 102 pupils are trained in different branches at the cost of its founder, a Sannyāsi depending on alms received from the public; and the Madhubanī Sanskrit school, which receives a Government grant of Rs. 20 a month, and prepares candidates for the first and second Sanskrit examinations and the Title examination. The latter has 3 Pandits on its staff, the cost of the establishment being Rs. 125 a month, of which Rs. 105 is met from subscriptions. The majority of the students of the schools are Brāhmans, and all of them are free students.

HOSTELS AND BOARDING HOUSES.

There are 3 boarding-houses in the district—one attached to the High English School at Madhubanī, which was opened in September 1905, the cost of the building being met from local subscriptions; the second is attached to the Madhubanī Sanskrit School; and the third to the Madrasa Imdādia in Darbhangā.

MUHAMMADAN EDUCATION.

In the year 1885 the number of Muhammadans in all classes of schools was 8,667, and their percentage to the total number of scholars was 22·2. With the general decline in the total number of pupils of all denominations, which, as already mentioned in the paragraph on Primary Education, occurred between 1885 and 1895, the number of Muhammadan pupils decreased in 1895 to 4,930, and their percentage to the total number of scholars fell to 16·6. When in 1905 the attendance of pupils of all denominations increased, the number of Muhammadan pupils increased to 7,497, but the percentage remained almost the same, viz., 16·5. As a rule, secular instruction is not very popular among the Muhammadans of Bengal, and they are generally slower than the Hindus to grasp the advantages of education; but considering that the percentage of Muhammadans to the total population is only 12·11, it would appear that in Darbhangā they are more eager than the Hindus to give their sons a modern education.

FEMALE EDUCATION.

There was no girls' school in the district prior to 1883, when 3 schools with 37 pupils were entered in the returns; but 1,923 girls were returned in 1882 as reading in boys' schools, this number being the result of the introduction of a system under which rewards were given to teachers in boys' schools for teaching girls. In 1885 the number of girls reading in boys' schools was 1,702, the stipends to girls' schools varying from Rs. 3 to Rs. 5 a month, and the total expenditure on female education from the primary grant being Rs. 1,259. In 1895 the number of schools rose to 31 and of pupils to 446, but the number of girls reading in boys schools declined to 226, a decrease which may be attributed to the reasons already mentioned in the paragraph on Primary Education. In 1905 the number of girls' schools rose to 67 and that of pupils to 1,237, while the number of girls reading in boys' schools increased to 1,469, so that the total number of girls reading in public institutions rose from 672 in 1895 to 2,706 in 1905, or by more than 400 per cent. At the same time, the expenditure from public sources on female education rose from Rs. 465 to Rs. 2,309.

Of the 2,706 girls receiving instruction in public institutions, 2 were in the Upper Primary stage, 859 in the higher, and 1,845 in the lower section of the Lower Primary schools. There is no prejudice against male teachers being employed in girls' schools, provided that they belong to the same village, nor is there any objection to boys and girls of tender years reading together.

There is one Model girls' school at Dighrā in the Samāstipur subdivision, which had 28 pupils on the rolls on the 31st March 1905. Out of 34 Lower Primary Scholarships allotted in this district, one is reserved for girls.

NUMBER OF LITERATES.

The preceding account shows that education, though still backward, has made considerable progress in recent years, and this is corroborated by the statistics of the census of 1901. At this census a special return was made of those able to read and write any language, and people of whatever age who could satisfy this simple test were entered as literate, and those who could not as illiterate. In Darbhangā it was ascertained that only 1 female per mille could read and write, but on the other hand the percentage of literate males was 7·1, as compared with 5·8 in 1891 and 3·3 in 1881. The percentage of literates was however much lower than in the Province as a whole, where the percentage of males able to read and write was found to be 10·4, while 5 out of every 1,000 females were returned as literate.

CHAPTER XV.

GAZETTEER.

Ahiāri.—A village in the north-west of the head-quarters subdivision, situated a little to the south-east of Kamtaul, about 15 miles north-west of Darbhangā, close to the northern boundary of the subdivision. Here a religious gathering, called the Ahalyāsthān or Singheswarsthān, is held every March. It lasts for a whole day, and is attended by about 10,000 people, who first cleanse themselves in the holy waters of the *kund* or reservoir of Deokalī in *pargana* Tirsat, and then come barefoot to see the footprint of Sīta, imprinted on a flat stone, which is enshrined in a temple called the Ahalyāsthān. The story runs that in this place was the shrine of Gautama Rishi, whose wife Ahalyā was remarkable for her beauty. The sage was accustomed to rise early and bathe in the Ganges, 50 miles distant. One morning, when he had gone off sooner than usual, Indra, king of the gods, assuming Gautama's appearance and manner, told Ahalyā that it was still early, and that he would not start till later. Ahalyā admitted him, unconscious of the deception practised on her. When the sage returned, he cursed them both. Ahalyā became a stone, in which shape she remained for a thousand years, until Rāma on his way to Janakpur touched the stone, when she was forthwith translated to heaven. The village contains a handsome temple, which is visible for several miles round.

Baherā.—A large village in the head-quarters subdivision, situated about 20 miles south-east of Darbhangā. It is the head-quarters of a police-station and a trade centre of some local importance; but the main interest attaching to it is that it was the original head-quarters of the Darbhangā subdivision, which was constituted in 1845. Owing to its unhealthy and inconvenient situation, the offices were removed to Darbhangā in 1865.

Balrājpur.—A village in the Madhubanī subdivision, situated about 10 miles east of Khajaulī and 16 miles north-east of Madhubanī. The village contains the remains of a fort known as the *garh* or fort of Rājā Bal. Only the ramparts are standing, but these are in good repair. The walls, which enclosed a rectangle about

500 by 300 yards in length, are composed of large well-burnt bricks, nearly a foot long, and are still 10 feet in height; and the remains of round towers, about 40 feet in height, are standing at each of the four corners. The interior of the fort is overgrown with jungle and is uninhabited, except by a solitary *bairāgī*, who is in charge of the shrine of Rājā Bal. The latter is worshipped as a local divinity, but the people round have no knowledge as to who Rājā Bal was or when the fort was built. They believe, however, that Rāja Bal and his army still inhabit it, and in consequence of this they are afraid to bring the site into cultivation. They allow their cattle to graze there in the day time, but nothing would induce them to visit it at night. This popular superstition is doubtless the reason for the excellent state of preservation of the ramparts, for otherwise the people could certainly have removed the bricks of which it is composed. A former Subdivisional Officer, while carrying out some excavations in the fort, was attacked by fever and had to give up the work. This circumstance has greatly strengthened the superstition with regard to the sanctity of the site, as the people believe that Rājā Bal himself sent the fever. The country round is dotted with mounds, in which bricks similar to those in the fort are found. These may have been constructed at the same time as the fort to serve as outposts.

Bāsdeopur.—A village in the Madhubanī subdivision, situated about 10 miles to the east of Madhubanī. Its old name was Sankarpur, which was first changed to Sankarpur Gandhwār, and then to the present name. The traditional explanation of these changes is as follows:—There were two brothers—one called Gandh, the other Bhaur—both exceedingly powerful. Though nominally subject to the Rājā of Tirhut, each acquired a large territory, the lands owned by Gandh lying to the east of the Tiljūgā, and those owned by Bhaur to the south of the Karai. The Rājā tried hard to get rid of them, but was unsuccessful, until he induced two strangers to kill them. The strangers, after killing them, obtained their estates, the slayer of Gandh and his descendants thus acquiring the name of Gandhmāriyā (Gandh-killing), and the family of the slayer of Bhaur the name of Bhaursoriyā (Bhaur-subduing). The latter lived at Singiā in *pargana* Chak Mani, close to the Karai, and the former in Sankarpur, which then became Sankarpur Gandhwār, and was ultimately annexed to the Darbhangā Rāj. When Mahārājā Chhattar Singh of Darbhangā married, this village was given to his wife as a dowry; and she in her turn gave it to her second son Bāsdeo Singh. On the death of Chhattar Singh in 1839, his eldest son, Rudar Singh, succeeded to the Rāj, and *pargana* Jarail was made over to Bāsdeo Singh for his maintenance ; but as he claimed half of the Rāj and contested Rudar Singh's right, he refused to compromise himself by accepting Jarail, and retired to Sankarpur Gandhwār, his mother's gift. Here he erected temples and dug several tanks ; and the village, again changing its name, was called after him Bāsdeopur.

Berautpur.—A village in the Khajaulī thāna, 12 miles north of Madhubani, containing a *garh* or fort which covers an area of about two-thirds of an acre. There are traces of walls and rooms, and an idol of Mahādeo, represented by a cylindrical pillar standing in a hollowed-out recess. The fort is said to have been built by Rājā Beraut, whose date is placed about the time of the Mahābhārata. The Telis or oilmen of the place claim him as one of their caste, on the ground that the pillar alluded to was intended to represent an oil-mill, and to be an emblem of his caste.

Bhawāra.—A large village about half a mile to the south of Madhubanī. To the south are pointed out the remains of a fort or *garh*, which at one time had brick walls. The whole is now in ruins, and there is nothing in its external appearance to distinguish it from an ordinary tank. It is said to have been built by Raghu Singh, one of the early members of the Darbhangā Rāj family, who resided here till about 1762, when Pratāb Singh removed his household to Darbhangā. There are no temples of any importance or interest in the village ; but a mosque, now in ruins, with only a front wall and six arched doors remaining, is pointed out as having been built by Alā-ud-dīn Husain, one of the last independent kings of Bengal (1493—1518).

Darbhangā Rāj.—The largest estate in the district, owned by the Mahārājā of Darbhangā. The Darbhangā family traces its origin to one Mahesh Thākur, who is said to have come from Jubbulpore about the beginning of the 16th century. He took service as a priest with the descendants of Rājā Siva Singh, who still exercised a nominal supremacy in Tirhut, but when they collapsed before the advancing Muhammadan power, Mahesh Thākur managed to induce Akbar to grant him what are now the Darbhangā Rāj estates. He and his descendants gradually consolidated the power of the family; and about 1700 the title of Rājā was for the first time conferred on its head, Raghu Singh, by Alī Vardi Khān. He was given a *mukarari* lease of *Sarkār* Tirhut at an annual rent of a lakh of rupees; but soon afterwards Alī Vardi Khān, hearing of the enormous profits he made, seized his property and carried off his family as prisoners to Patna. The Rājā surrendered himself, and was ordered back to Tirhut as a mere revenue collector, a few villages and a grant of *sadui* (2 per cent. on the collections) being given to him on condition "that he should do justice and relieve distress; that he should put the country in a flourishing state and keep it so; that he should supply the ryots with the necessaries for cultivation, and be equally answerable to Government for the revenue collected through his dependants as for those immediately under his own superintendence."

Rājā Narendra Singh was the proprietor of the estate, when the British took possession of Bihār. On his death in 1760, he was succeeded by his adopted son, Pratāb Singh, who in 1762 removed the family residence to Darbhangā from Bhawāra near Madhubanī. The latter was succeeded in 1776 by his brother Madhu Singh, who was constantly at variance with the Collector owing to his refusal to engage for the revenue, until certain claims of his were acknowledged. He obstinately refused all terms offered to him, and did his best, by intimidation, to prevent farmers from taking the settlement of his estates or from collecting rents if they did take the settlement of them. This was a serious obstacle in the way of effecting a settlement of the district; and the Collector, incensed by his contumacy, was continually complaining of his insolence and obstructiveness. Referring to the terms of the grant made to Raghu Singh, he angrily declared that "instead of relieving distress, he has multiplied it; instead of doing justice he has been the instrument of injustice; instead of putting the country in a flourishing state, he has desolated it; instead of furnishing the ryots with the necessaries for cultivation, he has drove them by repeated extortions from the fields which once were loaded with the fruits of their industry."

Madhu Singh did not finally come to terms till after 1800, when he agreed to a comparatively fair *jamā* of 1½ lakh; and his estates, which had been let out in farm, were then restored to him. On his death in 1808, he was succeeded by his son Chhattar Singh, who rendered good service to Government at the time of the Nepāl war in 1814-15. He was the first to receive the title of Mahārājā, and since that time the title, though never formally recognized as hereditary, has been granted by Government as a personal distinction to each successive proprietor of the estate. In 1839, a few days before his death, he made over his estate to his eldest son, Rudar Singh, with the exception of a few villages which he gave to his younger son, Bāsdeo Singh. The latter then claimed half of the Rāj, alleging that the disposition of the property ought not to be regulated by the *kulāchār* or family custom, but by Hindu law. The Sessions Judge, whose decision was upheld by the High Court, held that Rudar Singh was entitled to the Rāj, and that Bāsdeo Singh was only entitled to *pargana* Jarail as maintenance, on the ground that the succession to the family property is regulated by the family custom ; that the eldest son succeeds to the Rāj, the younger sons obtaining sufficient landed property for their maintenance; and that the Rāj pays the Government revenue direct for them, and they reimburse it, the lands being assigned on condition that, failing male issue, they revert to the Rāj. This decision, which has settled once for all that the estate is impartible and that inheritance to it is regulated by primogeniture, has been instrumental in preserving the estate in its integrity, and in making its proprietor the greatest land-owner in Bihār. But the immediate effect of the litigation, combined with mismanagement, was to involve the estate in serious difficulties; so that when Mahārājā Maheswar Singh died in 1866 and the Court of Wards took charge of it for his minor son, it was 70 lakhs in debt and the revenue was only 16 lakhs. Under the management of the Court of Ward, the finances of the estate recovered, and 20 years later, when it was made over to the late Mahārājā Lakshmeswar Singh, it was in a flourishing condition.

The Rāj estates at present comprise lands situated in the districts of Darbhangā, Muzaffarpur, Gayā, Monghyr, Purnea and Bhāgalpur, amounting to over 2,400 square miles. The Mahārājā is also the owner of house property in the towns of Darbhangā, Muzaffarpur, Patna, Benares, Calcutta, Allahābād, Darjeeling and Simlā, and of the indigo concerns of Saraiyā and Bachaur in the Muzaffarpur district, Pandaul in Darbhangā and Gondwāra in Purnea. The principal residence of the Mahārājā is the Anandbāgh palace; a palace for the Rānī is being built at Darbhangā, and another is in process of completion at Rājnagar. The rent-roll exceeds 32 lakhs of rupees, and the Government revenue, including cesses, is 7¼ lakhs. The present system of management was introduced when the estate was under the Court of Wards, and is very complete. The estate is divided into circles of from fifty to two hundred villages each; each circle is in charge of a sub-manager, who is responsible to the Mahārājā for its efficient working, and under each sub-manager there are usually several *tahsīldārs* or rent-collectors in charge of groups of villages.

The present proprietor of the estate is Mahārājā Bahādur Sir Rāmeswar Singh, K.C.I.E. He was born in 1860, and in 1878 was appointed to the Statutory Civil Service; he served in several Bihār districts as an Assistant Magistrate, but resigned the service in 1885. He succeeded to the Rāj in December 1898, when his brother died heirless, and was made a Mahārājā Bahādur three weeks afterwards. In 1900 he was given the Kaisar-i-Hind gold medal; in 1902 he was made K.C.I.E.; and in 1904 he was elected by the non-official members of the Bengal Legislative Council as their representative on the Governor-General's Legislative Council.

Darbhangā subdivision.—Head-quarters subdivision of the district, lying between 25° 38′ and 26° 26′ N., and 85° 41′ and 86° 44′ E., and extending over 1,224 square miles. It is bounded on the north by the subdivision of Madhubanī, on the south by the Samāstipur subdivision, on the west by Muzaffarpur, and on the east partly by the Supaul subdivision of the Bhāgalpur District and partly by the Begusarai subdivision of the Monghyr District. Like the Madhubanī subdivision, it resembles in shape a fairly well defined parallelogram, its length from north to south being greater than its breadth from west to east. The greater part of the area which it comprises consists of a low-lying plain, containing a number of marshes and intersected by many streams and rivers, of which the most important are the Bāghmati, Little Bāghmati, Kamlā and Tiljūgā. The Bāghmati, which enters this subdivision from Muzaffarpur, forms a natural boundary between it and the Samāstipur subdivision, pursuing a south-easterly course till it empties itself into the Burh Gandak near Ruserā. The Little Bāghmati enters the subdivision from the Madhubanī subdivision near Pāli and runs past the town of Darbhangā down to Hayā Ghāt, where it is joined by the Bāghmati proper. The Karai, which traverses the whole of Ruserā thāna, issues from the Kamlā and the Bāghmati, and joins the Tiljūgā at Tilkeswar. In itself, it is an unimportant stream, but it is fed by the Bāghmatī, which cuts into it east of Hayā Ghāt. The Kamlā enters the subdivision at Singar Pandaul, and flowing east of Darbhangā joins the Tiljūgā at the south-eastern corner of Ruserā thāna. The Tiljūgā skirts the eastern boundary of the subdivision.

The subdivision, which was established in 1845, comprises the thānas of Darbhangā, Baherā and Ruserā and the police outposts of Darbhangā town, Jalā, and Singiā. Its population, which was 1,048,806 in 1891, rose to 1,065,595 in 1901, and the density is 871 persons to the square mile. It contains 1,306 villages and 2 towns, Ruserā and Darbhangā; its head-quarters are at Laheriā Sarai, which is included within the limits of the Darbhangā Municipality.

Darbhangā town.—Head-quarters station and principal town of the district, situated on the bank of the Little Bāghmati river. It probably takes its name from one Darbhangī Khān, a Muhammadan free-booter, of whom little or nothing is known, though local patriotism insists that the name is a corruption of *Dar-i-Bangal, i. e.*, the gate of Bengal. The whole country round the town becomes a swamp during the rains, being subject to inundations from the Kamlā and Little Bāghmati; and the civil station and public offices were therefore moved in 1884 to the suburb of Laheriā Sarai at the extreme south of the town.

The town stretches for five or six miles along the bank of the Little Bāghmati, which is spanned by two large iron girder bridges. Its most conspicuous feature is a number of large tanks, which give it a picturesque appearance during the rains. The names of the largest are Gangā Sāgar, Dīghī, Harāhi and Lakshmī Sāgar. The three first are situated in a line, with a drive passing from one to the other, and their united length is 6,000 feet; Harāhi being 1,600 feet long and 1,000 feet broad; Dīghī 2,400 feet long and 1,200 feet broad; the Gangā Sāgar 2,000 feet long and 1,000 feet broad. It has been conjectured that these were excavated in order to secure raised ground for soldiers' quarters, this theory resting on the belief that the town was once a Musalmān cantonment. The local tradition, however, is that in the time of Rājā Siva Singh, a fisherwoman, with a basket of fish on her head, was on her way to market, accompanied by her daughter-in-law. A kite from a neighbouring tree pounced down and carried away a fish from the basket. Instead of sympathising with her, the daughter-in-law began to laugh. Enraged at her unfilial conduct, the mother-in-law gave vent to her anger, and a hot quarrel ensued. All this was witnessed by the Rājā as he sat at his window, and he lost no time in sending for the women. He asked the younger woman the cause of her unseasonable laughter; but she begged hard to be excused, saying that if she told her story it would be certain death to her. The Rājā's curiosity being roused, he insisted on hearing her reason. "In the reign of king Yudhisthir," said the younger fisherwoman, "I was a kite. During the war of the Mahābhārata, I carried away the arm of a woman, with a golden bracelet weighing 80 maunds, and brought it here and ate it. I laughed at the thought of the petty greed of the puny kites of the present time, who do not mind pouncing down on a paltry fish." Saying this, she expired. The Mahārājā, curious to find out the truth of the story, ordered a series of tanks to be dug in the places pointed out. At last, his perseverance was rewarded by finding the skeleton of the arm, as well as the golden bracelet; and so the tank in which they were found was called *Harāhi* or the bone tank.*

In respect of population, Darbhangā ranks third among the towns of the Patna Division. The population of the town increased from 53,744 in 1872 to 65,955 in 1881 and to 73,561 in 1891, but fell again to 66,244 in 1901, the total number of inhabitants including 47,946 Hindus, 18,122 Muhammadans and 171 Christians. This decrease, however, is to a great extent fictitious, as the population of the town was abnormally large in 1891 on account of the presence of some 5,000 Brāhmans who had come to partake of a feast given by the Mahārājā of Darbhangā, whereas in 1901 the date of the census was an auspicious one for weddings, and consequently a large number of persons were temporarily absent. Communications by road are good in all directions. The town is connected with the North Ganges railway system by a line from Samāstipur on the south, which branches off at Darbhangā in two directions, the first north-west to Sītāmarhi and the second north-east to Khanwā Ghāt on the Kosi. A considerable trade is carried on, the principal exports being oil-seeds, *ghī* and timber; and the imports consisting of food-grains, salt, gunny cloth, piece-goods, lime and iron.

Darbhangā was constituted a municipality in 1864, and the area within municipal limits is 7 square miles. A considerable portion of the town is of a rural character, for which very little is required in the way of drainage beyond seeing that the drains are kept clear and that the rain-water is carried off quickly. The natural fall is in most cases from west to east, but some parts drain towards the west. Besides the three large tanks to the east of the town, which have already been mentioned, there are numerous smaller tanks, 300 or 400 in number, which at present receive the rain-water and some of the sullage water of the town. A drainage scheme for carrying off rain-water and having six main outfalls was prepared some years ago, but it is only in the crowded portion of the town that a scheme of shallow surface drains is required. The water-supply is obtained from tanks, wells and the river; there are 383 private wells and 39 public wells; and 3 tanks, one of which belongs to the municipality, have been reserved for drinking-water. The municipality contains 14 miles of metalled and 20 miles of unmetalled roads, and two markets, one belonging to the Darbhangā Rāj and the other to the municipality. A large portion of the town has sprung up in

* Names of Places in Bihār by Mr. J. Christian, Calcutta Review, 1891.

recent times; and it contains few buildings of any interest. The principal buildings are the residence of the Mahārājā of Darbhangā, a large modern building with extensive grounds covering 55 acres; the Darbhangā Rāj Hospital, which is one of the finest in Bihar; and the Victoria Memorial Hall, which has taken the shape of a Town Hall, at Laheriā Sarai, one wing being used by the Heycock Club and containing a library.

Jaynagar.—A village in the Madhubanī subdivision, situated a few miles south of the Nepāl frontier and a little east of the river Kamlā, in 26° 35′ N., and 86° 9′ E. Population (1901) 3,551. The village contains the remains of a mud fort now in ruins. It is surrounded by a moat, and the remains of several turrets on each of its four sides are still visible. There is a tradition that a Muhammadan general selected this place as the site for a fort to resist the incursions of the hill tribes, but finding a dead body in the ground, he considered the spot unlucky and abandoned it. It is probably one of the line of forts which Alā-ud-dīn Husain, King of Bengal (1493—1518), constructed from Kāmrūp in Assam to Bettiah, in order to resist the inroads of the hill tribes. Near the fort is an encampment made by the English during the Nepalese war.

Jhanjhārpur.—A village in the Madhubanī subdivision, situated on the Bengal and North-Western Railway about 14 miles south-east of Madhubanī, in 26° 16′ N. and 86° 17′ E. Population (1901) 5,639. It is a large and thriving village, noted for the brass utensils turned out by the local braziers, particularly the *pānbatta* or box for holding betel-leaf and the *gangājalī* or water-pot. The place is also noteworthy from the fact that all the children of the Mahārājā of Darbhangā are born there and not at Darbhangā. It seems that many of the Darbhangā family had died childless, and that Pratāb Singh, the Rājā of Darbhangā from 1760 to 1776, alarmed at this, consulted a *mahanth* who lived at the village of Murnam not far off. The *mahanth*, proceeding to Jhanjhārpur, burnt a lock of his hair, and said that whoever dwelt there would have a male issue. Pratāb Singh immediately commenced to build a house on the spot; but unfortunately for the *mahanth's* prediction, he died without issue before the house was finished. Madhu Singh, his brother, however, completed it and sometimes stayed there. The village formerly belonged to a family of Rājputs, but, as the Mahārānī always resided there when pregnant, Mahārājā Chhattar Singh (1808—1839) bought it from them. It contains a temple of Rakatmātā, and two bazars, one called Pratābganj from Pratāb Singh, and the other Srīganj from Madhu Singh's sister-in-law.

Jiwachh.—A village in the head-quarters subdivision, situated 3 miles north-east of Darbhangā on the banks of the Kamlā. Here a *melā* or religious gathering is held during full moon in the months of Kārtik (October-November) and Māgh (January-February). Very little trade is carried on at the fair, which is primarily a bathing festival; it is attended principally by women, who imagine that bathing in the Kamlā is a cure for sterility, and that it will give sickly children a new lease of life.

Kakraul.—A village in the Madhubanī subdivision, about 5 miles north-west of Madhubanī and on the Jaynagar road. The place is a centre of the weaving of *khokti* cloth which is of good and firm texture, and is largely bought by the Nepalese. It is also celebrated among the local Hindu community from the fact that one of its hamlets, Husainpur, contains the temple of Kapileswar Mahādeo, which, according to tradition, was built by one of the ancient sages, Kapila Muni, who is said to have lived in the village and to have placed an image of Siva in the temple. A fair is held there annually in January or February, when cloth, brass vessels and grain are sold.

Laheriā Sarai.—Suburb of Darbhangā town. *See* Darbhangā town.

Lehrā.—A village in the head-quarters subdivision, situated 13 miles east of Darbhangā on the main road from Madhubanī to Baherā. Tradition connects this place with Siva Singh, the most famous of the last Hindu kings of Tirhut, and ascribes to him the three great tanks in the village. The largest is known as Ghordaur, and legend relates that it was excavated by Siva Singh, who, mounted on horseback, fixed its boundary by holding a pitcher of water in his right hand and allowing the water to flow through a spout, as he galloped his horse at its utmost speed, until the water from the vessel had all been expended. This tank is about 2 miles long, but there is water only at one end, the rest being now under cultivation. It seems that an old bed of the Kamlā cut into it, and drained off nearly all the water. Siva Singh is said to have lived near this tank; and there are about 13 *bīghas* of land now covered with bricks and jungle, which are pointed out as the site of his palace.

Madhubanī subdivision.—Northern subdivision of the district, lying between 26° 2′ and 26° 40′ N. and 85° 45′ and 86° 44′ E., and extending over 1,346 square miles. It is bounded on the north by the Nepāl *Tarai*, on the west by the Sitamarhi subdivision of the Muzaffarpur District, on the east by the Supaul subdivision of the Bhāgalpur District, and on the south by the head-quarters subdivision of Darbhangā.

In shape, the subdivision resembles a parallelogram, its mean breadth from west to east being a little greater than its mean length from north to south. It consists of a rich alluvial plain intersected by numerous rivers and streams issuing from the Nepāl hills and running almost parallel to each other from north to south. The chief of these rivers are the Little Bāghmati, Kamlā, Karai, Balān and Tiljūgā. The Little Bāghmati enters the subdivision at Bishnupur Ugarpatī about 10 miles west of Benīpatī, and after being joined by its tributary the Dhaus near Bankutā in the same thāna, takes a south-easterly course. The Kamlā flows southward from the hills and falls into the Little Bāghmati near Dudhail. This river frequently changes its course, and its old beds are found all over the north of the subdivision; at present an independent stream, called the Kamlā proper, flows past Singar Paudaul. To the east of this river are the Little Balān, the Bhāti Balān or the Balān proper, and the Tiljūgā, which skirts the eastern boundary of the subdivision. The Little Batān is a deep narrow river with a well-defined bed running south through the eastern part of the Khajaulī and Madhubanī thānas. The Bhāti Balān is a wide river with a shifting sandy bed, liable to heavy floods but remaining practically dry during the greater part of the year. Like the Kamlā, it frequently changes its course, and its old beds are found all over the north of Phulparās thāna.

The subdivision, which was established in 1866, comprises the thānas of Madhubanī, Benīpatī, Khajaulī and Phulparās, and the police outposts of Jhanjhārpur, Madhwārpur, Deodhā, Ladaniā and Madhupur, besides beat-houses at Harlākhi and Lamkhā. The population rose from 1,014,700 in 1891 to 1,094,379 in 1901, of whom 965,352 were Hindus and 128,992 were Muhammadans. It is less densely inhabited than the rest of the district, there being 813 persons to the square mile; and it is the only subdivision where there is much room for further expansion. It contains many ridges of uplands suitable for *rabi* cultivation; but the staple crop is winter rice, and the paddy of Alāpur, Jabdi, and Bachaur is famous all over Bihār. It contains one town, Madhubanī, its head-quarters, and 1,084 villages. Narahiā is an important centre of the Nepalese grain traffic; at Jhanjhārpur, on the railway, brass utensils of a superior quality are manufactured; at Saurāth an important annual *melā* or religious festival is held; and Jaynagar is the site of an old fort.

Madhubanī town.—Head-quarters town of the subdivision of the same name, situated about 16 miles north-east of Darbhanga town in 26° 21′ N. and 86° 5′ E. Population (1901) 17,802.

Madhubanī is an important trading centre on the road from the Sakrī railway station to the Nepāl frontier. It was constituted a municipality in 1869, and the area within municipal limits is 4 square miles. The town runs north and south, the principal thoroughfare being in that direction. It contains the usual public offices, a dispensary, and a sub-jail with accommodation for 12 prisoners. To the north of the town is the residence of some large land-owners, known as the Madhubanī Bābus, who are connections of the Mahārājā of Darbhangā, as they are descended from Kīrat Singh, the third son of Madhu Singh, who was the proprietor of the Darbhangā Rāj at the end of the 18th century. The name Madhubanī means the honey-forest, and points to a time when the land on which the town stands was covered by an uncleared forest, from which bee's-wax was obtained by a scanty population.

Mālīnagar.—A village in the south-western corner of the head-quarters subdivision, situated on the north bank of the Little Gandak, on the road from Darbhangā to Pūsa, opposite the latter place. The village contains a temple of Mahādeo, which was built in 1844 by a local *baniyā*, one of whose family was given the title of Rai Bahādur for his loyalty during the Mutiny. His descendant has the right of selecting the priest, who is not paid a fixed salary, but receives rice and vegetables daily, besides the offerings made to the idol. At the time the temple was built, it was decided to hold a *melā*, called Rāmnāmi, in honour of Rāma; and since that time an annual *melā*, lasting from the 1st to the 5th April, has taken place in the village. It is partly religious, partly commercial, and is attended by 2,000 to 4,000 persons.

Mithilā.—The ancient name for the tract of country bounded on the north by the Himālayas, on the east by the river Kosi, on the south by the Ganges, and on the west by the Gandak. It comprised the present districts of Champāran, Muzaffarpur and Darbhangā, parts of the districts of Monghyr, Bhāgalpur and Purnea, and the strip of Nepāl *Tarai* lying between these districts and the lower ranges of the Himālayas.

Narahiā.—Village in the north-eastern corner of the Madhubanī subdivision situated a few miles north-east of Phulparās in 26° 22′ N. and 86° 32′ E. Population (1901) 5,069. Narahiā is connected by a road with the railway, and is an important centre for the Nepalese grain traffic.

Pūsa.—Village in the Samāstipur subdivision, situated near the right bank of the Little Gandak and near the western boundary of the district in 25° 59′ N., and 85° 40′ E. Population

(1901) 4,570. The village was acquired by Government in 1796; and other waste lands appertaining to Bakhtiyārpur, a village on the other side of the river (population 1,384) were assigned to Government in 1798 without any additional rent. Pūsa was long used as a stud depôt, but all stud operations were closed in 1874. The property was then for a short time utilized as a model or experimental farm, special attention being devoted to the growing and curing of tobacco; but the management was unsuccessful, and in 1877 it was leased out to Messrs. Begg, Dunlop & Co. for tobacco-growing experiments. The latter continued to hold the farm up to 1897, but, the firm having finally decided to abandon their experiments, the lease was not renewed. In 1904 the estate was made over to the Government of India as the site for an Imperial Agricultural College, research laboratory, experimental cultivation farm and cattle-breeding farm; the buildings required for this purpose are now being erected.

The estate covers an area of 1,280 acres, of which about 800 acres are culturable, the remainder being occupied by roads, avenues, house-sites, etc. It is 5 miles from Wainī station on the Tirhut State Railway, to which there is a good road, and it is within easy reach by road of the towns of Darbhangā and Muzaffarpur. The river Gandak, which is navigable for boats of moderate size, forms its boundary on the north and east, and affords water for irrigation, while its navigability forms a practical advantage of some value. The area of the estate may be roughly divided into tracts of three distinct classes:—(1) The central upland tract of about 500 acres, upon which the buildings, the avenues and most of the roads are situated, including about 350 acres of culturable land. (2) The southern low-lying tract of about 350 acres suitable for cultivation of rice and *rabi* crops. (3) The riverain tract of about 150 acres, extending along the bank of the river Gandak. This tract is usually flooded in the rains and is chiefly suitable for grass-cutting and grazing. It is proposed to use 400 acres, or if necessary a larger area, for cultivation; and the remainder will provide grazing for a herd of 200 cattle, which it is intended to establish for the improvement of local breeds.

The estate is reported to be unusually well adapted for the purposes of an experimental farm; for the conditions of soil and climate are such that many of the more important field and garden crops can be grown as dry crops; and with irrigation (for which the river which adjoins the estate affords special facilities, and which is also possible from wells) practically every important crop known in India can be grown. As regards the cattle farm, the estate is said to be admirably adapted for breeding; it is proposed to maintain the farm for the purpose of improving the local breeds, rather than of introducing foreign strains. It will be a head-quarters from which bulls can suitably be supplied to the districts of Bengal (other than the rice tracts) or to the eastern districts of the United Provinces. The Research Institute owes its inception to the generosity of Mr. Henry Phipps, who placed at the disposal of the Viceroy, Lord Curzon, a donation of £30,000 to be used in promoting scientific research. This institute. to which Mr. Phipps' name is to be given, will adjoin the farm, and will contain a library, museum, herbarium, such laboratories and offices as are necessary, a large central hall, and one or two lecture-rooms for the students of the Agricultural College which is to be established as part of the scheme. The Agricultural College is intended not only to provide for the needs of Bengal, but also to serve as a model for, and raise the standard of, agricultural colleges in other Provinces, and to provide for a more complete and efficient agricultural education than is now possible in any of the existing institutions. The complete course of instruction will be one of 5 years, and the curriculum will include agriculture (theory of practice), chemistry (inorganic, organic and agricultural), systematic and cryptogamic botany, agricultural entomology, geology as applied to agriculture, elementary physics and medicines in their application to agriculture. The practical side of the training will include veterinary science, so far as it is required by agriculturists, land survey and mensuration, farm management, farm accounts and allied subjects.

Buserā.—Town in the head-quarters subdivision, situated on the east bank of the Little Gandak just below the confluence of that river with the Bāghmati, in 25° 45′ N. and 86° 2′ E. Population (1901) 10,245. Buserā, owing to its position on the Little Gandak, was at one time the largest market in the south of the district, but, though it is still an important bazar, it has somewhat lost its importance since the opening of the railway. Buserā was constituted a municipality in 1869, and the area within municipal limits is 2·5 square miles.

Samāstipur subdivision.—Southern subdivision of the district, lying between 25° 28′ and 26° 5′ N. and 85° 31′ and 86° 1′ E., and extending over 778 square miles. It is bounded on the north by the Bāghmati river, which separates it from the head-quarters subdivision, on the west by the Hājīpur subdivision of Muzaffarpur, on the south by the Ganges, and on the east by the Begusarai subdivision of the Monghyr District. Physically, it resembles the subdivisions of Hājīpur and Begusarai, the whole tract forming a rich alluvial country noted for the fertility of its soil and for the excellence of its *rabi* crops. The Ganges skirts the subdivision on the south for 20 miles only; and the most important rivers traversing the subdivision are the Little or Burh Gandak, the Bayā, and the Jhamwāri and Balān, which are both offshoots of the Little Gandak. The Burh Gandak enters the subdivision from the west near Pusā, and then pursues a south-easterly course till it leaves the District near Buserā; throughout this portion of its course it marks a natural boundary, the uplands, which constitute the greater part of the subdivision, lying to the south and the marshy *doāb* of Wārisnagar to the north. The Bayā flows through a portion of the Dalsingh Sarai thāna and joins the Ganges below Dhanespur at the extreme south-east corner of the subdivision. The Jhamwāri and the Balān are branches of the Burh Gandak, which they leave at Pusā, and after flowing through the south-west of Samāstipur rejoin the parent stream in Monghyr.

With the exception of part of the *doāb* between the Bāghmati and Little Gandak rivers, the subdivision consists of a large block of upland interspersed with a few *chaurs* or marshes. It is the richest and most fertile part of the district, producing all the most valuable *rabi* and *bhadoi* crops, and it is the centre of the indigo industry in the Darbhangā District. Vegetation is luxuriant, and mango groves and clusters of bamboos abound, giving a pleasing variety to the landscape. Its population rose from 738,449 in 1891 to 752,637 in 1901, when there were 967 persons to the square mile, the density of population being greater than in any other part of the district. It contains one town, Samāstipur, its head-quarters, and 843 villages. The subdivision was formerly known as the Tājpur subdivision, which was established in 1867; and it comprises the thānas of Samāstipur (formerly Tājpur), Dalsingh Sarai and Wārisnagar (formerly Nāgarbastī), and the police outposts of Mohiuddīnnagar and Tājpur.

Samāstipur town.—Head-quarters town of the subdivision of the same name, situated on the south bank of the Little Gandak river in 25° 52′ N. and 85° 48′ E. Population (1901) 9,101. Samāstipur is an important junction on the Bengal and North-Western Railway, and is the site of railway workshops which employ 1,000 hands. The town extends a short distance along the river bank, and is a large commercial centre, a considerable trade being carried on in rice, oil-seeds, pulses, saltpetre and piece-goods. It was constituted a municipality in 1897, and the area within municipal limits is 1·75 square miles. The town contains the usual public offices, and a sub-jail with accommodation for 23 prisoners.

Saurāth.—A village in the Madhubanī subdivision, situated 8 miles north-west of Madhubanī in 26° 24′ N. and 86° 31′ E. Population (1901) 2,062. The village is the site of a large *melā*, or religious gathering, which takes place annually in June or July, when thousands of Brāhmans assemble to settle their children's marriages. It contains a large temple of Mahādeo, which was built in 1845 by the Mahārājā of Darbhangā, a plain brick building, covered with stucco, of no great architectural pretensions. Close by is a large tank with broad steps leading down to the water, the whole place being shaded by a fine mango grove, under which the people buy and sell during the *melā*. Saurāth contains two large mounds, with ruins scattered for about a mile round, which the villagers believe to be the remains of an ancient city. Mr. Garrick of the Archæological Survey of India considered this belief well founded; and on making some small excavations in one of the mounds, unearthed some ancient bricks and a number of clay balls with holes through the centre, which, in his opinion, may have been used for spinning weights. *See* Reports Arch. Surv. Ind., Vol. XVI, pp. 94—97.

Silānāth.—A village in the Madhubanī subdivision, situated on the banks of the Kamlā, close to Jaynagar. Here every March or April a fair is held for about fifteen days, attended by 15,000 people, principally from the *Tarai*. The fair, doubtless, had its origin in large bodies of pilgrims coming to visit a temple of Mahādeo, which stood here; but the river Kamlā has changed its course, and washed the temple away, and now no trace of it remains.

Singiā.—A village in the Samāstipur subdivision, situated about 20 miles north-east of Tuserā and 2 miles north of the Karai river. Two miles to the south of the river is an old fort known as Mangalgarh or Mangal's fort. This is a large enclosure about 1½ mile in circumference, surrounded by what are now mud walls, 30 or 40 feet high, and by a deep ditch. The interior is under cultivation; but the ground is strewn with large bricks, varying from 1½ to 2 feet in length, showing that there must have been considerable buildings inside. Little is known about the history of this fort, but there is a tradition that Rājā Bal, whose stronghold has already been mentioned (*see* Balrājpur), attacked it and destroyed Rājā Mangal after having blown down the gates.

Tājpur.—A village in the Samāstipur subdivision of Darbhangā district, Bengal, situated in 25° 52′ N. and 85° 41′ E. Population (1901) 1,240. It was formerly the head-quarters of the present Samāstipur subdivision, but was never of much importance, for, even when it contained the subdivisional offices and a Munsif's court, the principal inhabitants were only court officials, *mukhtārs*, etc.

Tirhut.—The ancient name for the tract of country bounded on the north by the Himālayas, on the south by the Ganges, on the west by the river Gandak and on the east by the river Kosi. It comprised the British districts of Champāran, Muzaffarpur and Darbhangā, as well as the strip of Nepāl *Tarai* which runs between these districts and the lower ranges of the Himālayas. The name is a corruption of Tīrabhukti, or the river-side land; and according to tradition, it means the land in which the three great mythical *homas* (sacrificial fires) were performed, one at the birth of Sīta in or near Sītāmarhi, the second at Dhanukha at the foot of the Himālayas when the great celestial bow of Hara was broken by Rāma, and the third at the capital of Mithilā, Janakpur (now in Nepāl), at the marriage of Sīta. General Cunningham* considers that the term referred to the lands lying in the valleys of the Little Gandak and Bāghmati rivers. He points out that all the chief places in the country are found upon the banks of the former river, which, he says, must have been the channel of the Great Gandak until the seventh century A. D. Others, again, are of opinion that the term means the province bordering on the Ganges, and as the word *bhukti* is often used in the Sena inscriptions in the sense of province during the 10th, 11th and 12th centuries, it has been conjectured that the term is not very ancient and that it was first used when the Sena kings of Bengal conquered the country and settled a number of Bengali Brāhmans in it.

The name Tīrabhukti appears, however, to be far more ancient, for in the excavations† carried out at Basārh (Vaisāli) in the cold weather of 1903-04, seals dating back to the fourth and fifth centuries were discovered, on which the name occurs. Some of these seals were attached to letters addressed to officers, who are distinctly defined as being in charge of Tīrabhukti; and besides this word, there is the simple Tīra, which, it is suggested, was the locality from which the name Tīrabhukti or Province of Tīra was derived. The place cannot however be identified.

Under the Mughal Empire Tirhut formed a *Sarkār* or division of the *Sūbah* or province of Bihār. It comprised a very large tract of country, being bounded on the north by the Nepāl territory, on the south by *Sarkārs* Hājīpur and Monghyr, and on the east by *Sarkārs* Monghyr and Purnea. In other words, it included the

* Reports Arch. Surv. Ind., Vol. XVI.

† Report Arch. Surv. Ind. 1903-04.

present districts of Muzaffarpur and Darbhangā (except the southern portion, which was included in *Ṣarkār* Hājīpur), nearly the whole of Bhāgalpur and a small portion of Monghyr. In the early days of British administration, it formed a huge district covering the present districts of Muzaffarpur and Darbhangā, a large part of Bhāgalpur, *parganas* Kashmār in the south-east corner of Sāran, and *pargana* Bhads Bhusari in the south-western corner of Monghyr. The head-quarters of this district remained at Muzaffarpur, but various transfers of territory were made until the area of Tirhut coincided with that of Muzaffarpur, and Darbhangā. It was divided into two in 1875, the western portion being constituted the district of Muzaffarpur, and the eastern and larger portion being formed into the district of Darbhangā.

INDEX.

A.

B.

E.

F.

G.

H.

N.

O.

P

R.

S.

T.